WORKBOOKS IN CHEMISTRY

SERIES EDITOR

STEPHEN K. SCOTT

WORKBOOKS IN CHEMISTRY

Beginning mathematics for chemistry
Stephen K. Scott

Beginning calculations in physical chemistry
Barry R. Johnson and Stephen K. Scott

Beginning organic chemistry 1
Graham L. Patrick

Beginning organic chemistry 2
Graham L. Patrick

WORKBOOKS IN CHEMISTRY

Beginning Calculations in Physical Chemistry

BARRY R. JOHNSON AND STEPHEN K. SCOTT

School of Chemistry
University of Leeds

OXFORD NEW YORK TOKYO
OXFORD UNIVERSITY PRESS

This book has been printed digitally and produced in a standard specification in order to ensure its continuing availability

OXFORD
UNIVERSITY PRESS

Great Clarendon Street, Oxford OX2 6DP

Oxford University Press is a department of the University of Oxford.
It furthers the University's objective of excellence in research, scholarship, and education by publishing worldwide in

Oxford New York

Auckland Cape Town Dar es Salaam Hong Kong Karachi
Kuala Lumpur Madrid Melbourne Mexico City Nairobi
New Delhi Shanghai Taipei Toronto
With offices in
Argentina Austria Brazil Chile Czech Republic France Greece
Guatemala Hungary Italy Japan South Korea Poland Portugal
Singapore Switzerland Thailand Turkey Ukraine Vietnam

Published in the United States
by Oxford University Press Inc., New York

Oxford is a registered trade mark of Oxford University Press in the UK and in certain other countries

Published in the United States
by Oxford University Press Inc., New York

Reprinted 2006

ISBN 0-19-855965-8

Workbooks in Chemistry: Series Preface

The new *Workbooks in Chemistry Series* is designed to provide support to students in their learning in areas that cannot be covered in great detail in formal courses. The format allows individual, self-paced study. Students can also work in groups guided by tutors. Teaching staff can monitor progress as the students complete the exercises in the text. The Workbooks aim to support the more traditional teaching methods such as lectures. The format of the Workbooks has been evolved through experience and discussions with students over several years. Students benefit through the Examples and Exercises that provide practice and build confidence. University staff faced with increasing class sizes may find the Workbooks helpful in encouraging 'self-learning' and meeting the individual needs of their students more efficiently. The topics covered in the early Workbooks in the Series will concentrate on background support appropriate to the early years of a Chemistry degree, including mathematics, performing calculations, and basic concepts in organic chemistry. These should also be of interest to students who are taking chemistry courses as part of other degree schemes, such as biochemistry and environmental sciences. Later Workbooks will be designed to support material typically encountered in later years of a Chemistry course.

Preface

This Workbook has grown out of our observations and discussions with students at various stages within the Chemistry degree scheme at Leeds. Many students comment on their lack of confidence when asked to tackle explicit quantitative calculations in physical chemistry. This produces a 'barrier' to the underlying theory and sets up something of a vicious circle in which their confidence is repeatedly undermined. The chemistry then gets a reputation as being too 'difficult' to attempt. Students even with apparently impressive A-level grades can find themselves in this situation.

The present Workbook represents our attempt to build confidence in our students in this particular skill, which we believe is a necessary part of any chemist's training. The Workbook is not a 'physical chemistry textbook' *per se* and does not attempt to introduce and teach the background theory. The early sections deal with material for which the background theory might reasonably be expected to have been covered by students prior to entry to University and also provides a guided demonstration of SI units, powers of 10 etc. which are usually 'assumed' at the university level. Later sections have a higher proportion of Exercises over Examples and begin to cover undergraduate level physical chemistry. We have also included a section on drawing and using graphs.

The Workbook is intended to be written in. At Leeds, students are introduced to the various sections at appropriate stages alongside their lectures. The students may choose to work in groups, combining a mixture of self-pacing and peer support with more formal monitoring by tutorial staff as the booklet is completed.

It is a pleasure to acknowledge the help of the various Chemistry undergraduates who have helped the material improve as they have gone through it, and more specifically Mrs Elin Carson and Dr Peter Laye for their comments at various stages.

Leeds
October 1996

B. R. Johnson
S. K. Scott

Contents

Values of some important physical constants

Avogadro constant	N_A	6.022×10^{23} mol^{-1}
speed of light	c	2.998×10^{8} m s^{-1}
Planck constant	h	6.626×10^{-34} J s
Gas constant	R	8.314 J K^{-1} mol^{-1}
Boltzmann constant	k_B	1.381×10^{-23} J K^{-1}
elementary charge	e	1.602×10^{-19} C
mass of electron	m_e	9.109×10^{-31} kg
mass of proton	m_p	1.673×10^{-27} kg
mass of neutron	m_n	1.675×10^{-27} kg
Faraday constant	F	96485 C mol^{-1}
electron volt	eV	96485 J mol^{-1}

Greek alphabet

Greek letter		Name	Occurrence
Lower case	Upper case		
α	Α	alpha	
β	Β	beta	
γ	Γ	gamma	
δ	Δ	delta	small or finite change in a quantity
ε	Ε	epsilon	molecular energy
ζ	Ζ	zeta	
η	Η	eta	
θ	Θ	theta	
ι	Ι	iota	
κ	Κ	kappa	
λ	Λ	lambda	wavelength
μ	Μ	mu	micron (10^{-6} m)
ν	Ν	nu	frequency, ($\bar{\nu}$, wavenumber)
ξ	Ξ	xi	
ο	Ο	omicron	
π	Π	pi	pi, product (maths)
ρ	Ρ	rho	density
σ	Σ	sigma	summation (maths)
τ	Τ	tau	
υ	Υ	upsilon	
φ	Φ	phi	
χ	Χ	chi	
ψ	Ψ	psi	wavefunction
ω	Ω	omega	

SECTION 1

Revision: Powers of 10

Revision: Powers of 10

Before starting with some actual calculations, it is worthwhile revising the way powers of 10 and the terms kilo, milli etc. are used in chemistry. A longer discussion of the 'problem' of units etc. will be presented in Section 3 and methods for converting between different units will be explained at various other points in this workbook. Whilst these various techniques are not particularly 'chemistry', they are often the cause of mistakes in calculations and hence cause frustration and undermine confidence.

1.1 10^x notation

Rather than write numbers with strings of zeros either before or after the decimal point, it is usual to find large or small numbers written as a 'normal-sized' number multiplied by an appropriate power of ten.

Example 1.1

the number 300 can be written as 3×10^2
the number of molecules in one mole of substance in 6.022×10^{23}
the mass of a proton is 1.673×10^{-27} kg

In each case, the power of 10 indicates how many times the number should be multiplied by 10:

$$10^1 = 10$$
$$10^2 = 10 \times 10 \ (= 100)$$
$$10^3 = 10 \times 10 \times 10 \ (= 1000)$$

In general

10^x denotes multiplication x-times by ten

A negative exponent, as in 10^{-3}, indicates *division* by the appropriate power of 10:

$$10^{-1} = \frac{1}{10} = 0.1$$
$$10^{-2} = \frac{1}{10^2} = \frac{1}{100} = 0.01$$
$$10^{-3} = \frac{1}{10^3} = \frac{1}{1000} = 0.001$$

In general

$10^{-x} = 1/10^x$

Exercise 1.1

Write out the full form for (i) the number of molecules in one mole of substance and (ii) the mass of a proton:

(a) 6.022×10^{23} =

(b) 1.673×10^{-27} kg =

(Answers: (a) 602 200 000 000 000 000 000 000; (b) 0.000 000 000 000 000 000 000 000 001 673 kg)

Exercise 1.2

Express the following numbers in their full form:

(a) 2.998×10^{8}

(b) 9.6485×10^{4}

(c) 9.109×10^{-31}

Express the following numbers in terms of the appropriate powers of 10:

(d) 0.000 000 000 000 000 000 000 000 000 000 000 662 6

(e) 101 325

(Answers: (a) 299 800 000; (b) 96 485; (c) 0.000 000 000 000 000 000 000 000 000 000 910 9; (d) 6.626×10^{-34}; (e) 1.01325×10^{5})

One other convention that it is sometimes useful to remember is that

$$10^0 = 1$$

so any number n can be written as $n = n \times 10^0$

1.2 Entering powers of 10 on a calculator

The number $2 \times 10^3 = 2000$ can be entered on a calculator using the EE or EXP key as

2 EXP 3

Similarly, the number $2 \times 10^{-3} = 0.002$ could be entered using the above recipe with the +/– key

2 EXP +/– 3

One common mistake occurs, however, when trying to enter a number such as 1×10^3 or 1×10^{-3}.

If these numbers are written out in these full forms, there is usually no problem, we simply enter 1 EXP 3 or 1 EXP +/– 3 respectively.

However, a shorthand form for numbers of the general form 1×10^x is commonly found in textbooks, in which the $1 \times$ part is omitted. The shorthand form uses the notation 10^x.

Example 1.2

We often find the definition of the Ångstrom unit written as

$$1\ \text{Å} = 10^{-10}\ \text{m}$$

The temptation is to enter this number on a calculator as

10 EXP +/– 10 **THIS IS WRONG!**

Exercise 1.3

Enter the above number in the form 10 EXP +/– 10 on a calculator and then press the = key.

The number should change to 1. –09, indicating that the number entered is actually 10^{-9}. This arises because the number entered in this way is 10×10^{-10} when we really wanted to enter 1×10^{-10}.

Reminder

The number written as 10^x is actually 1×10^x.

When entering such a factor into a calculator, make sure that you enter 1 EXP x rather than 10 EXP x (which is actually 10×10^x or 10^{x+1})

Example 1.3

1×10^{-3} should be entered as 1 EXP –3

Exercise 1.4

Use the following information to obtain the molar mass M of sucrose in kg mol^{-1}:

M(sucrose) = 342.3 g mol^{-1} and 1 g = 10^{-3} kg

(Answer: enter 342.3 × 1 EXP +/–3: M(sucrose) = 0.3423 kg mol^{-1})

1.3. Shifting the decimal point

The 10^x notation is useful if we need to move the position of the decimal point in a number.

Example 1.4

We can 'tidy up' numbers so that there is only one digit in front of the decimal point:

$$345 = 3.45 \times 10^2$$
$$0.0345 = 3.45 \times 10^{-2}$$

The rules governing the change in the *exponent* x in 10^x and the movement of the decimal point are:

Each time we move the decimal point one place to the *left*, we *increase* the exponent in the power from x to $x + 1$

Each time we move the decimal point one place to the *right*, we *decrease* the exponent in the power from x to $x - 1$

These rules applied in the previous example.

To show this we can proceed stage-by-stage with 345

(i) write 345.0 as 345.0×10^0 (see the convention at the end of Section 1.1)

(ii) shifting the decimal point one place to the left, we increase the power by one

$$345.0 \times 10^0 = 34.5 \times 10^1$$

(iii) repeating stage (ii) we have

$$34.5 \times 10^1 = 3.45 \times 10^2$$

The rules can be expressed in an alternative way:

Each time the exponent in the power of 10 is *increased* by 1, the number 'in front' has to be *divided* by 10
Each time the exponent in the power of 10 is *decreased* by 1, the number 'in front' has to be *multiplied* by 10

Numbers displayed in this form, i.e. with a power of 10, are often said to be written in *scientific format.*

Exercise 1.5

(a) Rewrite the number 0.0254 in scientific format by moving the decimal point two places to the right

0.0254 =

This process is used in converting lengths from metres to centimetres (see next sections)

(b) Rewrite the following numbers so that they all have the power 10^{-3} at the end:

(i) 0.0254 =
(ii) 0.00254 =
(iii) 25.4 =

(c) Rewrite the following numbers so each has only one digit before the decimal point:

(i) 12.7 =
(ii) 0.00127 =
(iii) 1270 =

(Answers: (a) 2.54×10^{-2}; (b) (i) 25.4×10^{-3}, (ii) 2.54×10^{-3}, (iii) 25400×10^{-3}; (c) (i) 1.27×10^{1}, 1.27×10^{-3}, (iii) 1.27×10^{3})

1.4 The SI system

The SI system (see Section 3) of notation emphasises the powers of 10^3 and 10^{-3} and multiples of these such as 10^6, 10^9, . . . 10^{-6}, 10^{-9} etc.

These various powers of 10 are each denoted by a special prefix and associated symbol; common powers in chemistry are given in the table below:

power	prefix	symbol	power	prefix	symbol
10^1	deca	da	10^{-1}	deci	d
10^2	hecto	h	10^{-2}	centi	c
10^3	kilo	k	10^{-3}	milli	m
10^6	mega	M	10^{-6}	micro	μ
10^9	giga	G	10^{-9}	nano	n
10^{12}	tera	T	10^{-12}	pico	p
			10^{-15}	femto	f

(The prefix and symbol for 10^{-1} and 10^{-2} are also included in the above table as these are relevant and occur relatively frequently in chemistry.)

Example 1.5

The bond length R in the molecule HCl is 0.000 000 000 127 450 m

Express this in terms of the above prefix symbols.

The value for R can be written as

$$R = 0.12745 \times 10^{-9}\,\text{m} \text{ or } 127.45 \times 10^{-12}\,\text{m}$$

so R = 0.12745 nm or R = 127.45 pm

There is some flexibility for personal choice here as quoting the result in either nanometres or picometres is equally valid: the latter is perhaps more common in the case of bond lengths although the unit of angstrom (1 Å = 10^{-10} m) is also used - the above result being 1.2745 Å

Exercise 1.6

Express the following quantities in an appropriate form based on the above prefixes:

(a) the enthalpy of formation of H_2O(g) at 298 K

$$\Delta_f H_m^{\theta}(H_2O,\ g,\ 298\,K) = -241\ 820\ J\ mol^{-1} =$$

(b) the enthalpy of combustion of propane at 298 K

$$\Delta_{comb}H^{\theta}(\text{propane, g, 298 K}) = -2\,220\,000 \text{ J mol}^{-1} =$$

(c) the wavelength of blue light

$$\lambda = 0.000\,000\,47 \text{ m} =$$

(Answers: (a) -241.82 kJ mol^{-1}; (b) -2.22 MJ mol^{-1}; (c) 470 nm or 0.47 μm)

1.5 Multiplying and dividing powers of 10

When multiplying powers of 10 together we can make use of the following rules:

When two powers are multiplied, their exponents are simply added together
When one number is divided by another, the exponents are subtracted

Example 1.6

(a) Multiply 3×10^5 by 2×10^3 (without the use of a calculator)

We can write this as

$$3 \times 2 \times 10^5 \times 10^3$$

as the order of multiplying numbers is not important. The first part $3 \times 2 = 6$ and the powers of 10 are multiplied using the above rule

$$10^5 \times 10^3 = 10^{(5+3)} = 10^8$$

so

$$3 \times 10^5 \times 2 \times 10^3 = 6 \times 10^8$$

(b) Divide 3×10^5 by 2×10^3

We can write this as

$$\frac{3}{2} \times \frac{10^5}{10^3} = \frac{3}{2} \times 10^5 \times 10^{-3} = 1.5 \times 10^{5-3} = 150$$

(c) Divide 3×10^{-2} by 6×10^{-8}

We write this as

$$\frac{3}{6} \times \frac{10^{-2}}{10^{-8}} = \frac{3}{6} \times 10^{-2} \times 10^8 = 0.5 \times 10^{-2+8} = 0.5 \times 10^6 = 5 \times 10^5$$

Notice in (c) that dividing a number by a quantity which is much smaller than 1 results in a large answer.

Always check that your final answer makes sense in light of the magnitude of the numbers you are manipulating, as simply missing a minus sign on entering numbers into a calculator can result in an answer many powers of 10 wrong.

Exercise 1.7

This exercise illustrates the 'reverse' of the above point. What is the wavelength of blue light in cm?

$$\lambda =$$

(Answer: we can begin by writing $\lambda = 4.7\times10^{-7}\,\text{m}$

Next, we can split the power 10^{-7} into two factors $10^{-5} \times 10^{-2}$ so

$$\lambda = 4.7\times10^{-5}\times10^{-2}\,\text{m} = 4.7\times10^{-5}\,\text{cm})$$

These rules are particularly useful in obtaining quick estimates of the approximate size (the technical phrase for this is *the order of magnitude*) of two or more numbers being multiplied or divided.

Example 1.7

In Section 4 we will often wish to convert the energy ε of one molecule into the corresponding molar energy E, i.e. the energy per mole. This involves multiplying by the Avagadro constant.

If $\varepsilon = 1\times10^{-19}\,\text{J}$ and $N_A = 6.022 \times 10^{23}\,\text{mol}^{-1}$, estimate the order of magnitude of $E = \varepsilon \times N_A$

We have $E = 1 \times 10^{-19}\,\text{J} \times 6.022 \times 10^{23}\,\text{mol}^{-1} = 6.022 \times 10^{23} \times 10^{-19}\,\text{J mol}^{-1}$

The factor $10^{23} \times 10^{-19} = 10^{23-19} = 10^4$, so we have

$$E = 6.022 \times 10^4\,\text{J mol}^{-1} = 60.22\,\text{kJ mol}^{-1}$$

Exercise 1.8

(a) The Gas constant R is given by

$$R = k_B \times N_A$$

where $k_B = 1.3807 \times 10^{-23}\,\text{J K}^{-1}$ is the Boltzmann constant. Using the information $6.022 \times 1.3807 = 8.314$, find the value of R without using a calculator

$$R =$$

(b) Estimate the order of magnitude of the following term for the energy ε

$$\varepsilon = \frac{6.626\times10^{-34} \times 2.998\times10^{8}}{2\times10^{-6}}\,\text{J}$$

(Hint: use the approximations $6.626 \approx 6.6$, $2.998 \approx 3$ and $6.6 \times 3 \approx 20$ to obtain your estimate)

What is the equivalent molar energy E?

$$E =$$

(c) The Faraday constant is the electrical charge on 1 mole of electrons and has a value of 96.485 kC mol^{-1}, where C is the Coulomb, the unit of electrical charge. What is the charge per gram of electrons given that the mass of a single electron is 9.109×10^{-31} kg ?

$$\text{Charge per gram of electrons} = \frac{\text{Charge per mole of electrons}}{\text{Number of electrons per mole}} \times \text{Number of electrons in 1 g}$$

(Hint: if your answer is incorrect by a factor of 10^3 or 10^6 you should re-examine the question remembering that the order of magnitude may also be 'hidden' in the units. If you are still unsure how to do this calculation return to it after you have studied Chapter 2)

(Answers: (a) $R = 1.3807 \times 10^{-23}$ J K^{-1} $\times$ 6.022×10^{23} mol^{-1} = $1.3807 \times 6.022 \times 10^{23-23}$ J K^{-1} mol^{-1} = 8.314×10^{0} J K^{-1} mol^{-1} = 8.314 J K^{-1} mol^{-1}; (b) $\varepsilon \approx (6.6 \times 3/2) \times 10^{-34+8-(-6)}$ J = $9.9 \times 10^{-34+8+6}$ J = 9.9×10^{-20} J $\approx 1 \times 10^{-19}$ J; $E = \varepsilon \times N_A \approx 60$ kJ mol^{-1}; (c) 176 MC g^{-1})

SECTION 2

Calculating Masses of Atoms and Molecules

Calculating Masses of Atoms and Molecules

2.1 Atomic masses and molar masses

There are many sources that give the *relative atomic mass* A_r for each element.

This section is concerned with using these data to calculate the corresponding masses of individual atoms.

We begin by recognising the relationship between *relative atomic mass* and the *molar mass* M of an element.

The formal definition of relative atomic mass is:

$$A_r = \frac{\text{mass of 1 atom of the element}}{\frac{1}{12} \times \text{mass of 1 atom of } ^{12}\text{C}}$$

Colloquially, the relative atomic mass is sometimes (but incorrectly) called 'the mass in grams of one mole of atoms of an element'. It is defined as a **ratio** of masses and hence is a number with no units. These numbers appear underneath element symbols in Periodic Tables and in many textbooks.

A related quantity is the *molar mass* M. This is simply the mass of a sample containing exactly one mole of a given atom (in this case).

The molar mass of ^{12}C is 12 g mol^{-1} or 0.012 kg mol^{-1}

Note that the molar mass has units and that these must always be included when we work with this quantity.

The molar mass is numerically equal to the relative atomic mass with the units g mol^{-1} appended.

[**Note**: The definition of the SI unit of a *mole* is based on the ^{12}C isotope: one mole is the *amount of substance* that contains as many 'entities' (i.e. atoms, molecules etc.) as there are atoms in 0.012 kg of ^{12}C. A related quantity is the Avogadro constant, $N_A = 6.022 \times 10^{23}$ mol^{-1}]

Exercise 2.1

What are the molar masses of the following atoms?

(a) H, (b) O, (c) F, (d) Na, (e) P, (f) Rn

(Answers: (a) 1.0078 g mol^{-1}; (b) 16 g mol^{-1}; (c) 19 g mol^{-1}; (d) 22.99 g mol^{-1}; (e) 30.97 g mol^{-1}, (f) 222 g mol^{-1} or 0.222 kg mol^{-1}.)

When we require the mass of a single atom for calculations, there are **two** parts of the process:

(a) we must convert M from a molar quantity to that for a single atom

and (b) we must express the result in kilograms.

The first stage involves dividing M by the Avogadro constant ($N_A = 6.022 \times 10^{23}$ mol^{-1})

The second stage, is not strictly necessary as the mass can be quoted correctly in g (see next section) but is highly recommended if the resulting atomic mass is to be used in subsequent calculations. This conversion involves multiplying the result by 10^{-3} to convert from g to kg. (This is a rather loose explanation at present: a more rigorous and logical account of such conversions will be developed in Section 3.)

Reminder

When entering 10^{-3} on your calculator, enter 1 EXP–03 (i.e. remember that the number is 1×10^{-3}) not 10 EXP–03 (which is 10^{-2}: see previous section)

Example 2.1

For a hydrogen atom:

$$A_r = 1.0078 \text{ and, hence } M = 1.0078 \text{ g mol}^{-1}$$

Therefore, the mass of a single H atom, m_H, is found as follows:

$$m_H = \frac{M(\mathrm{H})}{N_A} = \frac{1.0078\,\mathrm{g\,mol^{-1}}}{6.022 \times 10^{23}\,\mathrm{mol^{-1}}} = 1.674 \times 10^{-24}\,\mathrm{g} = 1.674 \times 10^{-27}\,\mathrm{kg}$$

Exercise 2.2

Calculate the masses of the atoms of He and Ne

Use: $A_r(\mathrm{He}) = 4$ and $A_r(\mathrm{Ne}) = 20$

(a) He

$$m_{He} = \frac{M(\mathrm{He})}{N_A} = \underline{\qquad\qquad} = \qquad \mathrm{g} = \qquad \mathrm{kg}$$

(b) Ne

$$m_{Ne} =$$

(Answers: $m_{He} = 6.64 \times 10^{-27}\,\mathrm{kg}$; $m_{Ne} = 3.32 \times 10^{-26}\,\mathrm{kg}$)

One additional source of mistakes in this sort of calculation arises when elements exist with a variety of isotopes. For instance, Cl exists naturally as a mixture of $^{35}\mathrm{Cl}$ (75.5%) and $^{37}\mathrm{Cl}$ (24.4%).

The relative atomic mass A_r reflects the average composition of one mole of Cl atoms.

When dealing with a single atom, however, this will have to be one isotope or the other, not a mixture. However, the isotope to be considered is usually identified explicitly in problems.

Example 2.2

For ^{35}Cl: $A_r = 35$

$$m_{^{35}\text{Cl}} = \frac{M(^{35}\text{Cl})}{N_A} = \frac{35\,\text{g mol}^{-1}}{6.022 \times 10^{23}\,\text{mol}^{-1}} = 5.81 \times 10^{-23}\,\text{g} = 5.81 \times 10^{-26}\,\text{kg}$$

For ^{37}Cl, $m_{^{37}\text{Cl}} = 6.14 \times 10^{-26}\,\text{g}$

Exercise 2.3

Calculate the masses of atoms for the two isotopes of uranium ^{235}U and ^{238}U.

Use: $A_r(^{235}\text{U}) = 235$ and $A_r(^{238}\text{U}) = 238$.

(a) ^{235}U

$$m_{^{235}\text{U}} = \frac{M(^{235}\text{U})}{N_A} = \frac{\qquad\qquad}{\qquad\qquad} =$$

(b) ^{238}U

$$m_{^{238}\text{U}} =$$

(Answers: $m_{^{235}\text{U}} = 3.90 \times 10^{-25}\,\text{kg}$; $m_{^{238}\text{U}} = 3.95 \times 10^{-25}\,\text{kg}$)

Isotope masses and their abundances for all the elements are available in the literature, but for some of the more commom elements the relative atomic masses are listed in the table below.

Species	A_r
e^-	0.000549
p^+	1.007276
n	1.008665
^{1}H	1.007825
^{2}H	2.014102
^{4}He	4.002603
^{12}C	12.000000
^{13}C	13.003354
^{14}N	14.003074
^{16}O	15.994915
^{32}S	31.9721
^{35}Cl	34.9688
^{37}Cl	36.9651
^{235}U	235.043940
^{238}U	238.050810

2.2 Binding energies

We can compare the mass of atoms with the individual masses of the protons and neutrons of which they are composed. A more precise calculation of the mass of a single ^{4}He atom yields a mass:

$$m_{^4\mathrm{He}} = 6.6466 \times 10^{-27} \text{ kg}$$

Using the data for m_p and m_n from the table at the beginning of this book, we can compare this with the mass of two protons and two neutrons:

$$2m_p + 2m_n = 6.6953 \times 10^{-27} \text{ kg}$$

The mass deficit,

$$\Delta m = 2m_p + 2m_n - m_{^4\mathrm{He}} = 4.87 \times 10^{-29} \text{ kg}$$

may seem small but is actually real and very significant. It is related to the 'binding energy' for the ^{4}He nucleus, which can be evaluated using Einstein's formula $\varepsilon = mc^2$. In this case,

$$\varepsilon_b = \Delta mc^2 = 4.37 \times 10^{-12} \text{ J}$$

If we multiply this by the Avogadro constant N_A, to obtain the binding energy per mole, E_b, we find that this is equivalent to an energy of approximately 2.6 GJ mol^{-1}.

Exercise 2.4

(a) The relative 'atomic' mass for electrons is 5.485×10^{-4}. Calculate the mass of an electron.

(b) Calculate the mass of an atom of S, indicating which isotope you have evaluated this for.

(c) Calculate the mass of a ^{12}C atom.

(d) Compare the mass of a ^{12}C atom with that of its constituent protons and neutrons and hence calculate the binding energy for this atom (and for a mole of such atoms).

(Answers: (a) $m_e = 9.1 \times 10^{-31}$ kg; (b) for ^{32}S, $m = 5.31 \times 10^{-26}$ kg; (c) for ^{12}C, $m = 1.99 \times 10^{-26}$ kg; (d) $\Delta m = 1.59 \times 10^{-28}$ kg, so $\varepsilon_b = 1.43 \times 10^{-11}$ J $E_b = 8.6$ GJ mol^{-1})

2.3 Molecular masses

The mass of an individual molecule is calculated from the relative molecular mass M_r through the molar mass M in exactly the same way as for atoms in the previous sections.

(a) we divide M by Avogadro's constant N_A
(b) we express the result in kilograms not grams

Example 2.3

The mass of one molecule of H_2O: we use

$A_r(H) = 1$, $A_r(O) = 16$ so $M_r(H_2O) = 18$ and $M = 18\ g\ mol^{-1}$

$$m_{H_2O} = \frac{M(H_2O)}{N_A} = \frac{18\ g\ mol^{-1}}{6.022 \times 10^{23}\ mol^{-1}} = 2.99 \times 10^{-23}\ g = 2.99 \times 10^{-26}\ kg$$

Exercise 2.5

Calculate the molecular mass of $H\,^{35}Cl$ and $D\,^{37}Cl$

(a) $H\,^{35}Cl$: $A_r(H) = 1$, $A_r(^{35}Cl) = 35$

so $M(H\,^{35}Cl) =$

then $m_{H^{35}Cl} = \frac{M(H\,^{35}Cl)}{N_A} = \frac{\qquad}{\qquad} =$

(b) $D\,^{37}Cl$: $A_r(D) =$, $A_r(^{37}Cl) =$

so

and $m_{D^{37}Cl} =$

(Answers: 6.0×10^{-26} kg; 6.5×10^{-26} kg)

Exercise 2.6

The mean speed of an ideal gas molecule depends on its mass m and the temperature T, according to the formula

$$\bar{c} = \left(\frac{8k_BT}{\pi m}\right)^{1/2}$$

where $k_B = 1.381 \times 10^{-23}$ J K^{-1} is the Boltzmann constant.

Calculate the mean speed of an $H^{35}Cl$ molecule at 300 K

The result of this calculation will have the SI units for $\bar{c}$, namely m s^{-1} (as we will prove in Section 3.3). We can convert the result into the more familiar units of miles per hour (m.p.h.) by noting that 1 mile = 1609 m and 1 hour = 3600 s

In performing this calculation, think carefully about whether to multiply or divide by each conversion factor: a mile is longer than a metre, so there will be fewer miles per second than metres per second. Conversely, an hour is longer than a second, so the molecule will travel further per hour than per second:

(Answers: $\bar{c}$ = 420.1 m s^{-1} = 940 m.p.h., roughly twice the speed of a jet aeroplane)

More interesting molecules

In each of the following cases, calculate (i) the relative molecular mass, (ii) the molar mass and (iii) the molecular mass:

(a) the stimulant caffeine has the molecular formula $C_8H_{10}O_2N_4$

(i) M_r =

(ii) M =

(iii) m =

(b) the explosive trinitrotoluene (TNT) has the molecular formula $C_7H_5O_6N_3$

(i) M_r =

(ii) M =

(iii) m =

(c) adenosine triphosphate (ATP) is important for energy storage in biochemical reactions and has the molecular formula $C_{10}H_{14}O_{13}N_5P_3$

(i) M_r =

(ii) M =

(iii) m =

(Answers: (a) $M_r = 194$, $M = 0.194$ kg mol^{-1}, $m = 3.222 \times 10^{-25}$ kg; (b) $M_r = 227$, $M = 0.227$ kg mol^{-1}, $m = 3.77 \times 10^{-25}$ kg; (c) $M_r = 505$, $M = 0.505$ kg mol^{-1}, $m = 8.39 \times 10^{-25}$ kg)

2.4 Amount of substance

The *amount of substance*, usually given the symbol n, is related to the concept of a mole and is frequently (but strictly incorrectly) called the 'number of moles' of a substance.

1 mole of substance contains as many molecules or atoms as there are atoms in exactly 12 g of ^{12}C

An important quantity in this regard is the Avogadro constant, N_A. (Note that it is not called 'Avogadro's number'). This is defined as

$$N_A = \text{number of molecules or atoms etc. per mole}$$

and has the value $$N_A = 6.022 \times 10^{23}\ \text{mol}^{-1}.$$

(Note that there are units associated with N_A as we have seen in the various examples above.)

If we have a sample of mass m containing N molecules of a substance (say H_2 for an example), then the amount of that substance is obtained by dividing that number by Avogadro's constant N_A:

$$n = \frac{\text{Number of molecules}}{\text{Number of molecules per mole}} = \frac{N}{N_A}$$

We can also re-express this equation in terms of the mass m of our sample and the mass of one mole (i.e. the molar mass, M) of that substance

$$n = \frac{\text{Mass of molecules in sample}}{\text{Molar mass}} = \frac{m}{M}$$

The unit of amount of substance is *moles* (symbol, mol).

Example 2.4

Calculate the amount of substance in a sample of H_2 gas of mass 1 g

Using the second equation above, with $M_{H_2} = 2 \text{ g mol}^{-1}$

$$n = \frac{m}{M_{H_2}} = \frac{1 \text{ g}}{2 \text{ g mol}^{-1}} = 0.5 \text{ mol}$$

We can also use this result to calculate the number of H_2 molecules in the sample:

$$N = n \times N_A = 0.5 \text{ mol} \times 6.022 \times 10^{23} \text{ mol}^{-1} = 3.011 \times 10^{23}$$

Note that the units cancel, leaving a number as the answer.

Exercise 2.7

The mass of a cube of pure iron, of side length 2 mm, is 62.88 mg. What is the density of iron expressed in units of (a) kg m^{-3}, and (b) g cm^{-3}? Check that your final answers are consistent.

Approach: First calculate the volume of the cube in m^3 and cm^3. Then make use of the relationship; density $= \frac{\text{mass}}{\text{volume}}$. To use this sucessfully, you must first convert the mass of the cube into the appropriate units, as requested in the question.

(a) 2 mm = m

Volume of cube $= (\quad \times 10 \quad \text{m})^3$
$= \quad \text{m}^3$

62.88 mg = $\times 10 \quad$ kg

$\therefore$ density $= \dfrac{\times 10 \quad \text{kg}}{\text{m}^3}$

density of iron =

(b) 2 mm = cm

Volume of cube $= (\quad \times 10 \quad \text{cm})^3$
$= \quad \text{cm}^3$

62.88 mg = $\times 10 \quad$ g

$\therefore$ density $= \dfrac{\times 10 \quad \text{g}}{\text{cm}^3}$

density of iron =

To check your answers for consistency, convert your answer for (b) into the units of (a) by calculating the number of cm^3 there are in 1 m^3

(Answers: (a) 7860 kg m^{-3}; (b) 7.86 g cm^{-3}. There are 1×10^6 cm^3 in 1 m^3. If you are still unsure, return to this calculation after completing Section 5.7)

2.5 Reduced masses

Another quantity related to the mass of the individual atoms in a molecule is the *reduced mass* μ. This quantity is important in the rotation and vibration of molecules.

For a diatomic such as HCl, the reduced mass is given by

$$\mu = \frac{m_1 m_2}{m_1 + m_2} \quad \text{or} \quad \frac{1}{\mu} = \frac{1}{m_1} + \frac{1}{m_2}$$

(these give the same result) where m_1 and m_2 are the masses of the individual atoms.

Reduced masses can be calculated from the relative atomic masses A_r, by first calculating m_1 and m_2 from Section 2.1 above.

However, we can also calculate μ directly from the values of A_r.

First, we calculate a 'molar' reduced mass;
then, we divide by Avogadro's constant and
then we remember to express the result in kilograms not grams:

$$\mu = \frac{A_r(1)A_r(2)}{A_r(1) + A_r(2)} \times \frac{10^{-3}}{6.022 \times 10^{23}} \text{kg}$$

where $A_r(1)$ and $A_r(2)$ are the relative atomic masses of the two elements. The factor of 10^{-3} kg arises from the division of the product of the two masses by their sum.

Example 2.5

(a) The reduced mass of $H^{35}Cl$: $A_r(H) = 1$, $A_r(^{35}Cl) = 35$

$$\mu = \frac{A_r(H)A_r(^{35}Cl)}{A_r(H) + A_r(^{35}Cl)} \times \frac{10^{-3}}{6.022 \times 10^{23}} \text{kg} = \frac{1 \times 35}{1 + 35} \times \frac{10^{-3}\,\text{kg}}{6.022 \times 10^{23}} = \frac{35}{36} \times \frac{10^{-3}\,\text{kg}}{6.022 \times 10^{23}}$$

$$= 1.61 \times 10^{-27}\,\text{kg}$$

(b) The reduced mass of H_2:

$$\mu = \frac{1 \times 1}{1 + 1} \times \frac{10^{-3}\,\text{kg}}{6.022 \times 10^{23}} = \frac{1}{2} \times \frac{10^{-3}\,\text{kg}}{6.022 \times 10^{23}} = 8.30 \times 10^{-28}\,\text{kg}$$

Exercise 2.8

Calculate the reduced masses of the following molecules: CO, HF, HBr and HI

Use $A_r(\text{H}) = 1$, $A_r(\text{C}) = 12$, $A_r(\text{O}) = 16$, $A_r(\text{F}) = 19$, $A_r(\text{Br}) = 79$, $A_r(\text{I}) = 127$

(a) CO

$$\mu_{\text{CO}} = \frac{\quad \times \quad}{\quad + \quad} \times \frac{10^{-3}\,\text{kg}}{6.022 \times 10^{23}} = \frac{\quad}{\quad} \times \frac{10^{-3}\,\text{kg}}{6.022 \times 10^{23}} =$$

(b) HF

(c) HBr

(d) HI

(Answers: (a) 1.14×10^{-26} kg; (b) 1.58×10^{-27} kg; (c) 1.64×10^{-27} kg; (d) 1.65×10^{-27} kg)

Some interesting facts about reduced masses

If a molecule consists of one light and one heavy atom, the reduced mass will be very close to the atomic mass of the light atom.

(Compare the values for HF, HCl, HBr and HI with that for $m_{\text{H}} = 1.67 \times 10^{-27}$ kg from Section 2.1)

For homonuclear diatomics such as H_2, Cl_2 and O_2, the reduced mass is simply equal to one half of the atomic mass.

e.g.

$$\mu_{\text{H}_2} = \frac{m_{\text{H}} m_{\text{H}}}{m_{\text{H}} + m_{\text{H}}} = \frac{m_{\text{H}}^2}{2m_{\text{H}}} = \tfrac{1}{2} m_{\text{H}}$$

and similarly for any molecule in which both atoms are identical.

Reduced masses arise in problems associated with the *relative* velocity of two molecules (e.g. in reaction kinetics) and in rotational and vibrational spectroscopy.

Exercise 2.9

(a) The relative velocity of two atoms depends on their reduced mass and on the temperature according to the formula

$$\bar{c}_{\text{rel}} = \left(\frac{8k_BT}{\pi\mu}\right)^{1/2}$$

Calculate $\bar{c}_{\text{rel}}$ for H and F atoms, using the previous result for μ_{HF}, at 500 K

(b) Show that the relative velocity of two H atoms is given by $4(k_BT/\pi m_{\text{H}})^{1/2}$

(c) The *moment of inertia I* of a diatomic molecule involves the reduced mass μ and the *bond distance* r_e, with

$$I = \mu r_e^2$$

Calculate I for HI for which $r_e = 160$ pm $= 160 \times 10^{-12}$ m

(d) The natural oscillator frequency ν_{osc} of a diatomic molecule depends on the reduced mass and the *bond force constant k* according to

$$\nu_{\text{osc}} = \frac{1}{2\pi}\sqrt{\frac{k}{\mu}}$$

Calculate ν_{osc} for HBr for which k = 411.5 N m^{-1}. (We will show in Section 3.3 that the units for ν_{osc} that emerge from this calculation are those of frequency, i.e. s^{-1} provided k and μ are in SI units.)

(Answers: (a) 3336 m s^{-1}; (b) we start by noting that $\mu_{\text{H}_2} = \frac{1}{2}m_{\text{H}_2}$, substituting this into the equation for the relative velocity gives $\bar{c}_{\text{rel}} = \left(16k_BT/\pi m_{\text{H}_2}\right)^{1/2} = 4\left(k_BT/\pi m_{\text{H}_2}\right)^{1/2}$; (c) 4.22 × 10^{-47} kg m^2; (d) 7.97 × 10^{13} s^{-1})

2.6 Momentum, kinetic energy and de Broglie wavelength

The *momentum* (which is often represented by the letter p) of an atom or molecule is calculated from the mass m and the velocity v

$$p = mv$$

The mass and velocity also determine the kinetic energy ε_k

$$\varepsilon_k = \tfrac{1}{2}mv^2$$

The kinetic energy can also be related to the momentum

$$\varepsilon_k = \frac{p^2}{2m}$$

Example 2.6

We will calculate some molecular velocities in Section 7, where we will find that light gas molecules typically move at speeds of 1 km s^{-1}. Calculate the momentum of an H_2O molecule with such a velocity.

We calculated earlier $m_{H_2O} = 2.99\times10^{-26}$ kg, so

$$p = mv = 2.99\times10^{-26}\,\text{kg}\times1000\,\text{m s}^{-1} = 2.99\times10^{-23}\,\text{kg m s}^{-1}$$

(Note that 1 km s^{-1} = 1000 m s^{-1})

The units for momentum can be written in other ways (as we will see in Section 3): for instance, the units are equivalent to J m^{-1} s which will be important when we calculate the de Broglie wavelength below.

Exercise 2.10

(a) Calculate the kinetic energy of the H_2O molecule in the above example.

(b) Calculate the momentum and kinetic energy of an electron with a velocity of 600 km s^{-1}

(Answers: (a) $\varepsilon_k = 1.5 \times 10^{-20}$ J; (b) $p = 5.47 \times 10^{-25}$ kg m s^{-1}, $\varepsilon_k = 1.64 \times 10^{-19}$ J)

The SI unit of kinetic energy is the joule, J.

One use of the momentum of a molecule is in the calculation of the *de Broglie wavelength* which arises in quantum mechanics.

The de Broglie wavelength is given by the formula

$$\lambda = \frac{h}{p} = \frac{h}{mv}$$

where h is the Planck constant, $h = 6.626 \times 10^{-34}$ J s

(Note that the units of h and the units of p, which can be written as J m^{-1} s, mean that λ will have units of m, i.e. length)

Example 2.7

The de Broglie wavelength corresponding to an H_2O molecule with velocity 1 km s^{-1} (for which $p = 2.99 \times 10^{-23}$ J m^{-1} s as calculated above) is given by

$$\lambda = \frac{h}{p} = \frac{6.626 \times 10^{-34}\,\text{J s}}{2.99 \times 10^{-23}\,\text{J m}^{-1}\,\text{s}} = 2.22 \times 10^{-11}\,\text{m} = 22.2\,\text{pm}$$

(We can compare this with the OH bond length $l = 96$ pm in the H_2O molecule.)

de Broglie wavelengths can be calculated for any bodies of any size that have momentum, not just molecules.

Exercise 2.11

Calculate the de Broglie wavelengths for the following:

(a) an electron with velocity 600 km s^{-1}

(b) a neutron with velocity 600 km s^{-1}

(c) a TNT molecule with velocity 200 m s^{-1} (using the mass calculated in Section 2.3)

(d) an athlete of $m = 75$ kg sprinting at 10 m s^{-1}

(e) a car of $m = 750$ kg travelling at 100 m s^{-1} (try to use the answer to (d) to find this value without actually calculating h/p)

(Answers: (a) $\lambda = 1.21$ nm; (b) $\lambda = 0.66$ pm; (c) $\lambda = 8.79$ pm; (d) $\lambda = 8.8 \times 10^{-37}$ m; (e) noting that both the mass and the velocity are a factor of 10 larger than in (d), the momentum will be 100 times larger and hence λ will be 100 times smaller, giving $\lambda = 8.8 \times 10^{-39}$ m. The latter two cases in particular have such short de Broglie wavelengths that quantum effects will not be evident.)

2.7 Moment of inertia and rotational constants

The reduced mass μ arises in several contexts, including its role in the *moment of inertia* which is important when considering the rotation of a diatomic molecule.

The moment of inertia I for a diatomic can be calculated from

$$I = \mu r_e^2$$

where μ is the reduced mass (Section 2.5) and r_e is the *equilibrium bond distance* between the atoms.

Example 2.8

The equilibrium bond distance for the HCl molecule $r_e = 127$ pm

Earlier we calculated $\mu_{HCl} = 1.61 \times 10^{-27}$ kg, so

$$I = 1.61 \times 10^{-27}\text{ kg} \times (127 \times 10^{-12}\text{ m})^2 = 2.60 \times 10^{-47}\text{ kg m}^2$$

This is a typical order of magnitude for the moment of inertia of a small diatomic molecule.

The moment of inertia is used to calculate the *rotational constant*, B, for a given molecule:

$$B = \frac{h}{8\pi^2 cI}$$

where c is the velocity of light.

Example 2.9

The rotational constant for HCl can be calculated from the information above

$$B_{HCl} = \frac{6.626 \times 10^{-34}\text{ s}}{8 \times \pi^2 \times 2.998 \times 10^8\text{ ms}^{-1} \times 2.60 \times 10^{-47}\text{ kg m}^2} = 1077\text{ m}^{-1}$$

The units for the rotational constant turn out to be those of inverse length. We will see in Section 4 that B is actually a *wavenumber*. The rotational constant can be determined directly from rotational or microwave spectroscopy. In a microwave spectrum, we see a series of *absorption lines* that have a wavenumber spacing of $2B$. We can use this information in the opposite direction and hence calculate the moment of inertia and the equilibrium bond length of diatomic molecules if we know their microwave spectrum.

Exercise 2.12

The microwave spectrum of CO shows a series of lines separated by 384.6 m^{-1}

What is the value of the rotational constant B_{CO}?

From this, calculate the moment of inertia I_{CO} and, using the reduced mass calculated in Section 2.5, find the equilibrium bond distance.

(Answers: B_{CO} = separation/2 = 192.3 m^{-1}; $I_{CO} = h/8\pi^2 cB_{CO} = 1.46 \times 10^{-46}$ kg m^2; r_e = 113 pm)

Important Equations used in this Section

The following should be familiar after completing this section and are gathered here for reference:

molar mass	$M = M_r \times 1\,\mathrm{g\,mol^{-1}} = \dfrac{M_r}{1000}\,\mathrm{kg\,mol^{-1}}$
molecular mass	$m = \dfrac{M}{N_A} = \dfrac{M_r}{1000 \times 6.022 \times 10^{23}}\,\mathrm{kg}$
amount of substance	$n = \dfrac{N}{N_A}$
reduced mass	$\mu = \dfrac{m_1 m_2}{m_1 + m_2}$
momentum	$p = mv$
kinetic energy	$\varepsilon_k = \frac{1}{2}mv^2 = \dfrac{p^2}{2m}$
de Broglie wavelength	$\lambda = \dfrac{h}{p}$
moment of inertia	$I = \mu r_e^2$
rotational constant	$B = \dfrac{h}{8\pi^2 cI}$

SECTION 3

Units: Dimensional Analysis

Units: Dimensional Analysis

3.1 Physical quantities, dimensions and units

Physical quantities, such as a reaction enthalpy or a boiling point, have two essential components:

a numerical value

and

an associated set of units

The units associated with a physical quantity depend on the *dimensions* of that quantity.

There are seven physical quantities that are described as being *dimensionally independent*:

dimension	*symbol*	SI unit name and symbol
length	l	metre, m
mass	m	kilogram, kg
time	t	second, s
temperature	T	kelvin, K
amount of substance	n	mole, mol
electric current	I	ampere, A
luminous intensity	I_v	candela, cd

Thus, we might measure a mass $m = 0.1$ kg or a length $l = 0.2$ m.

(Note that some care is needed to distinguish between the use of m as a symbol for the quantity mass in the first case and m as the unit metre for the second.)

Quantities that have different dimensions (sets of units) cannot be added, subtracted, or even compared. For instance, we cannot meaningfully say that a length of 10 m is larger or smaller than a temperature of 100 K as these relate to quite different things.

3.2 SI base units

The base units for the seven physically independent quantities under the SI system (le Systėme International d'Unités) are listed in the table above along with the recommended unit symbol. The definitions of these base units can be found in the book 'The International System of Units' by the National Physical Laboratory, published by Her Majesty's Stationery Office.

The units for the dimensions appropriate to all other physical quantities can be built from these base units, as indicated in the next section.

3.3 SI derived units

A number of additional SI units exist. These include the joule for energy, the newton for force and the watt for power. Each of these derived units can be expressed uniquely in terms of the base units.

Once the basic definitions have been established, many useful relationships between the derived units can also be recognised.

Some examples are given below: in each case, we begin from the definition of the quantity in physical terms based on the dimensions in the table on the previous page.

Example 3.1

Derive the definition of the SI unit for speed in terms of the base units.

The appropriate definition of speed is 'rate of change of distance';

the appropriate dimensions are, therefore, distance per unit time;

the appropriate SI unit is, therefore, m s^{-1}

Although it is defined slightly differently, velocity has the same units as speed.

Exercise 3.1

Acceleration, *a*, is defined as 'rate of change of velocity': determine the appropriate SI units for acceleration.

(Answer: the definition implies that the units are velocity per unit time: i.e. (m s^{-1}) s^{-1} = m s^{-2})

The SI units for area and volume are m^2 and m^3 respectively

For the next set of quantities, the following definitions are useful:

force = mass × acceleration
energy = force × distance
pressure = force/unit area

Example 3.2

From the previous definition, derive the SI unit for force.

The derived SI unit of force is the newton, N

thus force = mass × acceleration

$$\text{N} = \text{kg} \times \text{m s}^{-2}$$

The definition is thus, $\text{N} = \text{kg m s}^{-2}$

Exercise 3.2

Derive the SI unit for:

(a) energy, i.e. the joule

energy =
J

(b) pressure, i.e. the pascal (Pa)

pressure =
Pa

(Answers: (a) $J = N\ m = kg\ m^2\ s^{-2}$; (b) $Pa = N\ m^{-2} = kg\ m^{-1}\ s^{-2}$)

It is also of interest to compare the units for pressure and energy:

$$Pa = \frac{kg}{m\ s^{-2}} = \frac{kg\ m^2}{m^3\ s^{-2}} = \frac{J}{m^3}$$

i.e. pressure is dimensionally equivalent to energy per unit volume.

This result is useful when manipulating the ideal gas equation

$$pV = nRT,$$

as illustrated in the next example.

Example 3.3

Show that the ideal gas law is dimensionally consistent and hence find the units of pressure × volume.

We replace the physical quantities by their units, noting that R has units of $J\ K^{-1}\ mol^{-1}$:

$$Pa \times m^3 = mol \times J\ K^{-1}\ mol^{-1} \times K$$

Now, writing $J\ m^{-3}$ for Pa, we get

$$J\ m^{-3} \times m^3 = mol \times J\ K^{-1}\ mol^{-1} \times K$$

Cancelling m^{-3} with m^3 on the left-hand side, and mol with mol^{-1} and K^{-1} with K on the right-hand side, we find then that this reduces to J = J, i.e. both sides have the same dimensions.

The units for pressure × volume (and also of nRT) are thus J, i.e. the units of energy or work.

Exercise 3.3

Calculate the pressure exerted by 1 mol of an ideal gas at a temperature $T = 298$ K in a volume of 22.4×10^{-3} m^3

Use $p = nRT/V$ and write the units explicitly in the equation as you proceed, cancelling as appropriate ($R = 8.314$ J K^{-1} mol^{-1}):

$$p = \frac{nRT}{V} =$$

(Answer: $p = 1.106 \times 10^5$ Pa)

Exercise 3.4

The mean speed of an ideal gas depends on the mass m of the molecule and the temperature T as given by the formula

$$c = \left(\frac{3k_B T}{m}\right)^{1/2}$$

The Boltzmann constant has units of J K^{-1} whilst the mass m has units of kg.

Show that this formula is consistent with the dimensions of speed.

(Answer: (J K^{-1} × K/kg)$^{1/2}$ = (J/kg)$^{1/2}$: substituting in for J = kg m^2 s^{-2}, we have (kg m^2 s^{-2}/kg)$^{1/2}$ = (m^2 s^{-2})$^{1/2}$ = m s^{-1}, which is the correct set of units)

An alternative form of the equation is $c = (3RT/M)^{1/2}$ where R is the Gas constant (units = J K^{-1} mol^{-1}) and M is the molar mass (units = kg mol^{-1}). Show that this is also dimensionally consistent.

The units of power, the watt W, defined as rate of change of energy, are

$$\mathrm{W} = \mathrm{J\,s^{-1}} = \mathrm{kg\,m^2\,s^{-3}}$$

This result is useful in obtaining the definition of the derived units for electrical quantities in terms of the base units.

Example 3.4

Using power = voltage × current i.e. W = V × A

we obtain $\mathrm{V} = \mathrm{W\,A^{-1}} = \mathrm{J\,s^{-1}\,A^{-1}} = \mathrm{kg\,m^2\,s^{-3}\,A^{-1}}$

The derived unit and its symbol for a number of physical quantities relevant to physical chemistry are given below, along with 'useful' relationships between these and the definition in terms of the SI base units.

quantity	SI unit	symbol	relationship to other quantities	basic definition
concentration	mole per cubic metre	$mol\ m^{-3}$		$mol\ m^{-3}$
molality	mole per kilogram	$mol\ kg^{-1}$		$mol\ kg^{-1}$
frequency	hertz	Hz		s^{-1}
wavenumber	reciprocal metre	m^{-1}		m^{-1}
area	square metre	m^2		m^2
volume	cubic metre	m^3		m^3
speed velocity	metre per second	$m\ s^{-1}$		$m\ s^{-1}$
acceleration	metre per square second	$m\ s^{-2}$		$m\ s^{-2}$
force	newton	N	$J\ m^{-1}$	$kg\ m\ s^{-2}$
energy, enthalpy free energy	joule	J	N m	$kg\ m^2\ s^{-2}$
entropy	joule per kelvin	$J\ K^{-1}$		$kg\ m^{-2}\ s^{-2}\ K^{-1}$
molar free energy	joule per mole	$J\ mol^{-1}$		$kg\ m^2\ s^{-2}\ mol^{-1}$
molar entropy	joule per mole kelvin	$J\ mol^{-1}\ K^{-1}$		$kg\ m^2\ s^{-2}\ mol^{-1}\ K^{-1}$
pressure	pascal	Pa	$N\ m^{-2}$, $J\ m^{-3}$	$kg\ m^{-2}\ s^{-2}$
power	watt	W	$J\ s^{-1}$	$kg\ m^2\ s^{-3}$
charge	coulomb	C		A s
potential difference voltage	volt	V	$J\ s^{-1}\ A^{-1}$	$kg\ m^2\ s^{-3}\ A^{-1}$
resistance	ohm	Ω	$V\ A^{-1}$	$kg\ m^2\ s^{-3}\ A^{-2}$
conductance	siemens	S	$A\ V^{-1}$, Ω^{-1}	$kg^{-1}\ m^{-2}\ s^3\ A^2$
capacitance	farad	F	$A\ s\ V^{-1}$	$kg^{-1}\ m^{-2}\ s^4\ A^2$
density	kilogram per cubic metre	$kg\ m^{-3}$		$kg\ m^{-3}$
diffusion coefficient	square metre per second	$m^2\ s^{-1}$		$m^2\ s^{-1}$
heat capacity	joule per kelvin	$J\ K^{-1}$		$kg\ m^2\ s^{-2}\ K^{-1}$
molar heat capacity	joule per mole kelvin	$J\ mol^{-1}\ K^{-1}$		$kg\ m^2\ s^{-2}\ mol^{-1}\ K^{-1}$
specific heat capacity	joule per kilogram kelvin	$J\ kg^{-1}\ K^{-1}$		$kg\ m^2\ s^{-2}\ kg^{-1}\ K^{-1}$
thermal conductivity	watt per metre kelvin	$W\ m^{-1}\ K^{-1}$	$J\ s^{-1}\ m^{-1}\ K^{-1}$	$kg\ m\ s^{-3}\ K^{-1}$
surface tension	pascal metre	Pa m	$N\ m^{-1}$	$kg\ s^{-2}$
dynamic viscosity	pascal second	Pa s	$N\ s\ m^{-2}$	$kg\ m^{-1}\ s^{-1}$

3.4 Why bother?

There is an unfortunate tendency for the business of including and manipulating units to be seen as separate, and of a rather lower level of importance, from the real part of calculations, i.e. from the number crunching.

The 'units part' is often treated as an optional extra, with some vague guess at the appropriate units appended to the numerical results at the very last line of a calculation. Equally depressing, it also sometimes appears to be regarded as simply another set of traps that can be set by examiners as an excuse to rob students of marks they otherwise deserve!

In fact, keeping units in with the numbers, and being able to recognise when groups cancel with each other, can be a great help in keeping a calculation on the right track.

Example 3.5

We frequently need to find the ratio of the energy gap between two energy levels, $\Delta\varepsilon$, and the 'thermal energy' given by RT, where T is the temperature and R the Gas constant.

If the energy is to be derived from the frequency ν of an absorption line in a spectrum, then we use the Planck formula

$$\Delta\varepsilon = h\nu$$

where h is the Planck constant, $h = 6.626 \times 10^{-34}$ J s

By inspecting the units in this equation, we see that the resulting energy will have units of J s $\times$ s^{-1} (from the frequency), i.e. the energy will have the units of joules.

As an example, the IR spectrum of CO has an absorption line corresponding to a frequency $\nu = 6.42 \times 10^{13}$ s^{-1} so

$$\Delta E = 4.25 \times 10^{-20}\ \text{J}$$

We might now try to compare this with RT at, say 298 K:

$$\begin{aligned} RT &= 8.314\ \text{J K}^{-1}\ \text{mol}^{-1} \times 298\ \text{K} \\ &= 2478\ \text{J mol}^{-1} \end{aligned}$$

We can see, however, that $\Delta\varepsilon$ and RT have different units, differing by mol^{-1}, and so they cannot be directly compared at this stage.

If we were to use these particular values in evaluating the Boltzmann factor $e^{-\Delta\varepsilon/RT}$, we would obtain a result indistinguishable from unity, and which would be numerically wrong as well as dimensionally inconsistent.

With experience, the presence of the extra unit mol^{-1} can act as a strong hint that there is a factor related to the Avogadro constant N_A (which also has these units) missing from our approach.

In this case, the quotient $\Delta\varepsilon/RT$ tries to compare a *molecular energy*, $\Delta\varepsilon$, with a *molar energy* RT.

In order to allow the comparison, and to allow us subsequently to take the exponential, we must either 'convert' $\Delta\varepsilon$ to a molar quantity, by multiplying by N_A, or use the 'molecular thermal energy' k_BT, where k_B is the Boltzmann constant:

$$k_B = 1.381 \times 10^{-23}\ \text{J K}^{-1}\ \text{(note the absence of mol}^{-1}\text{ in this constant).}$$

The molar energy, $\Delta E = N_A \times \Delta\varepsilon = 25.6$ kJ mol^{-1}, so then

$$\frac{\Delta E}{RT} = \frac{25600\ \text{J mol}^{-1}}{2478\ \text{J mol}^{-1}} = 10.33$$

This result has no units, i.e. it is a pure number, and so we can take the exponential to calculate the Boltzmann factor:

$$e^{-\Delta E_m/RT} = e^{-10.33} = 3.26 \times 10^{-5}$$

a result significantly different from 1.

Exercise 3.5

Repeat the above calculation, comparing the molecular energy $\Delta\varepsilon$ with k_BT, showing that all the units cancel in this case too and that the same Boltzmann factor is obtained.

Keeping units in the calculation also helps prevent factors of 10^3, perhaps arising from an energy expressed in kJ mol^{-1}, from being forgotten.

Exercise 3.6

Calculate the free energy of formation of H_2O(g) at 298 K from the following information:

$$\Delta_f G_m = \Delta_f H_m - T\Delta_f S_m$$

$$\begin{aligned}\Delta_f H_m(\mathrm{H_2O, g, 298\,K}) &= -241.82\ \mathrm{kJ\ mol^{-1}}\\ \Delta_f S_m(\mathrm{H_2O, g, 298\,K}) &= 188.83\ \mathrm{J\ K^{-1}mol^{-1}}\end{aligned}$$

Note first, that $\Delta_f H_m$ and $\Delta_f S_m$ have different dimensions, but that $\Delta_f H_m$ and $T\Delta_f S_m$ have the same dimensions and so one can be subtracted from the other to obtain $\Delta_f G_m$ (which also then has the same dimensions).

Typically, $\Delta_f H_m$ values are quoted in kJ mol^{-1} as this is typically their magnitude.

Typically, $\Delta_f S_m$ are quoted in J K^{-1} mol^{-1}

(Answer:

$$\begin{aligned}\Delta_f G_m &= -241.82\ \mathrm{kJ\ mol^{-1}} - 298\ \mathrm{K} \times 188.83\ \mathrm{J\ K^{-1}\ mol^{-1}}\\ &= (-241\,820 - 298 \times 188.83)\ \mathrm{J\ mol^{-1}}\\ &= (-241\,820 - 56270)\ \mathrm{J\ mol^{-1}}\\ &= -298\,090\ \mathrm{J\ mol^{-1}}\\ &= -298.09\ \mathrm{kJ\ mol^{-1}}\end{aligned}$$

)

3.5 Guggenheim notation

The Guggenheim notation system is used widely (and ought to be adopted universally) to provide an unambiguous but compact way of representing sets of data so that these can be tabulated, plotted and compared 'at a glance'.

If we return to the beginning of this section, as mentioned, every physical quantity has an associated numerical value and a set of units. It may also be convenient to include a power of 10 raised to some exponent, so in general we may have something of the form

$$\text{value of quantity} = \text{numerical value} \times \text{power of } 10 \times \text{units}$$

For instance, we may have measured the frequency ν of an absorption line in an infra-red spectrum of a diatomic and obtained

$$\nu = 2.6 \times 10^{12} \text{ s}^{-1}.$$

We may, in fact, have been able to measure the frequencies of several lines in the spectrum, and we might wish to record these as well. Perhaps, the values of the next three lines are as follows:

$$\nu = 5.20 \times 10^{12} \text{ s}^{-1},$$
$$\nu = 7.80 \times 10^{12} \text{ s}^{-1},$$
$$\nu = 1.04 \times 10^{13} \text{ s}^{-1}.$$

If there is an extensive list, it is inconvenient to have to write the power of 10 and the unit along with the numerical value for each reading. A neater solution would be to have a column of numbers only, with the power of 10 and the unit associated in the column heading along with the symbol, ν in this case.

To achieve this, we can treat the three parts of the right-hand side of these 'equations' by the usual rules of algebra, just as if they were *x*s or *y*s in a traditional mathematical equation. Thus, we can divide both sides of the equation by the power of 10 and by the unit to give

$$\frac{\nu}{10^{12} \times \text{s}^{-1}} = 2.6$$

for the first reading.

Exercise 3.7

Repeat this process for the remaining readings, producing a result of the form above

(Answer: $\frac{\nu}{10^{12} \times \text{s}^{-1}} = 5.2,\ 7.8$ and $\frac{\nu}{10^{13} \times \text{s}^{-1}} = 1.04$)

The power of 10 is sometimes brought into the numerator, rather than leaving it in the denominator, so that we could write the first reading as

$$\frac{10^{-12}\nu}{\text{s}^{-1}} = 2.6$$

noting that $1/10^{12} = 10^{-12}$.

In order to tabulate the four readings, it is actually best to re-write all four so that they have the same power of 10, e.g. by re-writing the fourth reading as $\nu = 10.4 \times 10^{12}\ \mathrm{s}^{-1}$, before moving the units and power of 10 to the left-hand side. We can then produce a column, headed with the left-hand side, which consists otherwise of pure numbers:

$10^{-12}\ \nu\ /\mathrm{s}^{-1}$
2.6
5.2
7.8
10.4

The (not very well disguised) trend in these numbers is then immediately apparent.

If we find a value reported in Guggenheim notation, we can reverse the above process to obtain an individual value of the physical quantity.

Example 3.6

Determine the value of the rate constant k at a temperature of 298 K from the following table

T/K	$10^5\ k/\mathrm{mol}^{-1}\ \mathrm{dm}^3\ \mathrm{s}^{-1}$
293	1.77
298	2.46
303	3.38
308	5.03

The first column relates to the temperature, and the second to the rate constant. The second row has the following values associated with it:

$$\frac{T}{\mathrm{K}} = 298 \qquad \text{and} \qquad \frac{10^5 k}{\mathrm{mol}^{-1}\mathrm{dm}^3\mathrm{s}^{-1}} = 2.46$$

Multiplying the first equation through by the unit K, we have:

$$T = 298\ \mathrm{K}$$

For the data extracted from the second column, we divide through by 10^5 and multiply through by the units, to give

$$k = 2.46 \times 10^{-5}\ \mathrm{mol}^{-1}\ \mathrm{dm}^3\ \mathrm{s}^{-1}$$

an appropriate form for a second-order rate constant.

Because the two columns of data in the above example are now simply numbers, we can also use them to plot a graph. The axes should then be labelled with the same form as the column headings. This is shown in the following graph:

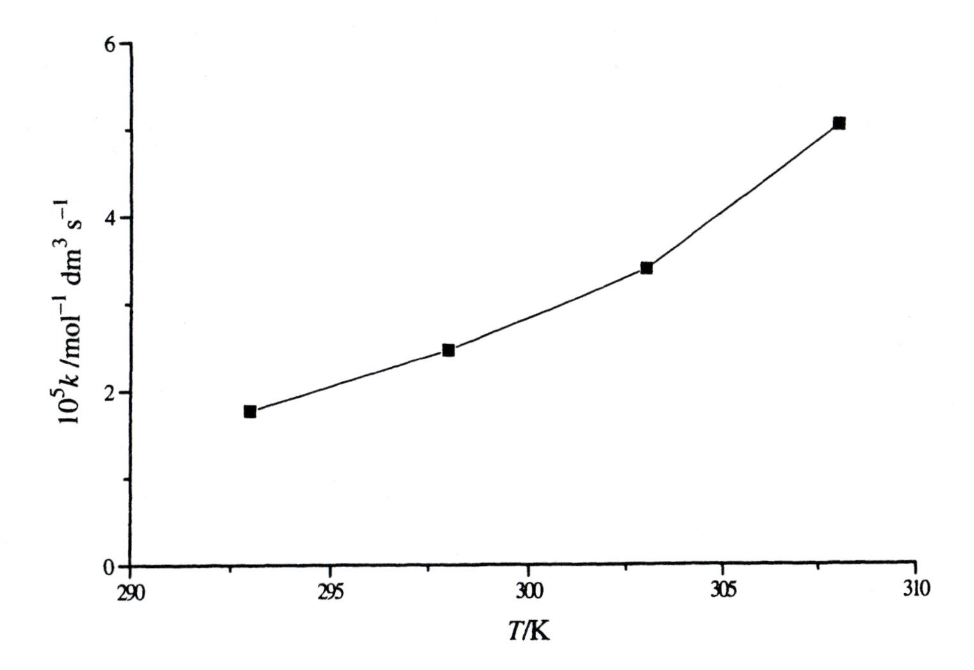

where the data points have been joined by a smooth curve.

The craft of plotting data in appropriate graphical form will be the subject of Section 6. We may note here, however, that a second graph from the rate constant data may be appropriate. In order to determine the *activation energy* for a reaction, it is usual to plot ln(k) versus the inverse temperature $1/T$.

Again, however, we must not ignore the units, and so should construct appropriately headed columns in our data table and use appropriately labelled axes on our graph.

Example 3.7

Beginning with the temperature, we have for the first row

$$T/\text{K} = 293 \qquad \text{implying } T = 293 \text{ K}$$

Using the second form, and inverting *both* sides of the equation, we have

$$\frac{1}{T} = \frac{1}{293} \times \frac{1}{\text{K}}, \quad \text{i.e.} \quad \frac{1}{T} = \frac{3.41 \times 10^{-3}}{\text{K}}$$

If we now use the Guggenheim method to move the unit and power of 10 to the left-hand side, we obtain

$$\frac{10^3 \text{K}}{T} = 3.41$$

For the rate constant, we take the logarithm of the number given in the column of data, so for ths first row we have to evaluate ln(1.77) = 0.571. The appropriate column heading must continue to indicate that we have units and a factor of 10^5 implicit in this number: 1.77 is equal to the value of 10^5 k/mol^{-1} dm^3 s^{-1} and so the heading should be ln {10^5 k/mol^{-1} dm^3 s^{-1}}

Exercise 3.8

Complete the following table with the appropriate column headings and entries

T/K	$10^5\ k$/mol^{-1} dm^3 s^{-1}		
293	1.77	3.41	0.571
298	2.46		
303	3.38		
308	5.03		

Plot the data and label the axes on the following graph:

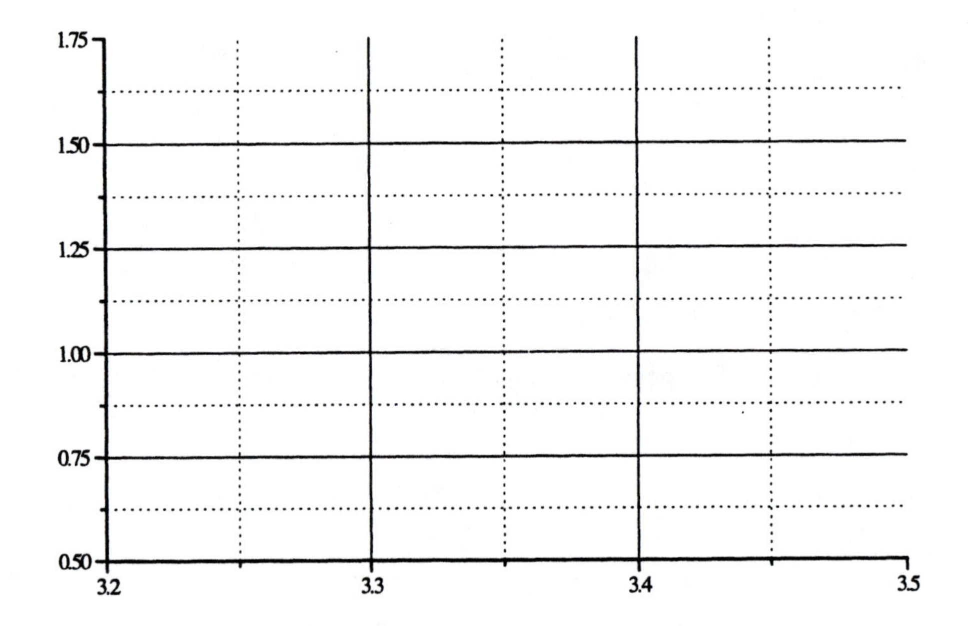

Answer

T/K	$10^5\ k$/mol^{-1} dm^3 s^{-1}	10^3K/T	ln{$10^5\ k$/mol^{-1} dm^3 s^{-1}}
293	1.77	3.41	0.571
298	2.46	3.36	0.900
303	3.38	3.30	1.218
308	5.03	3.25	1.615

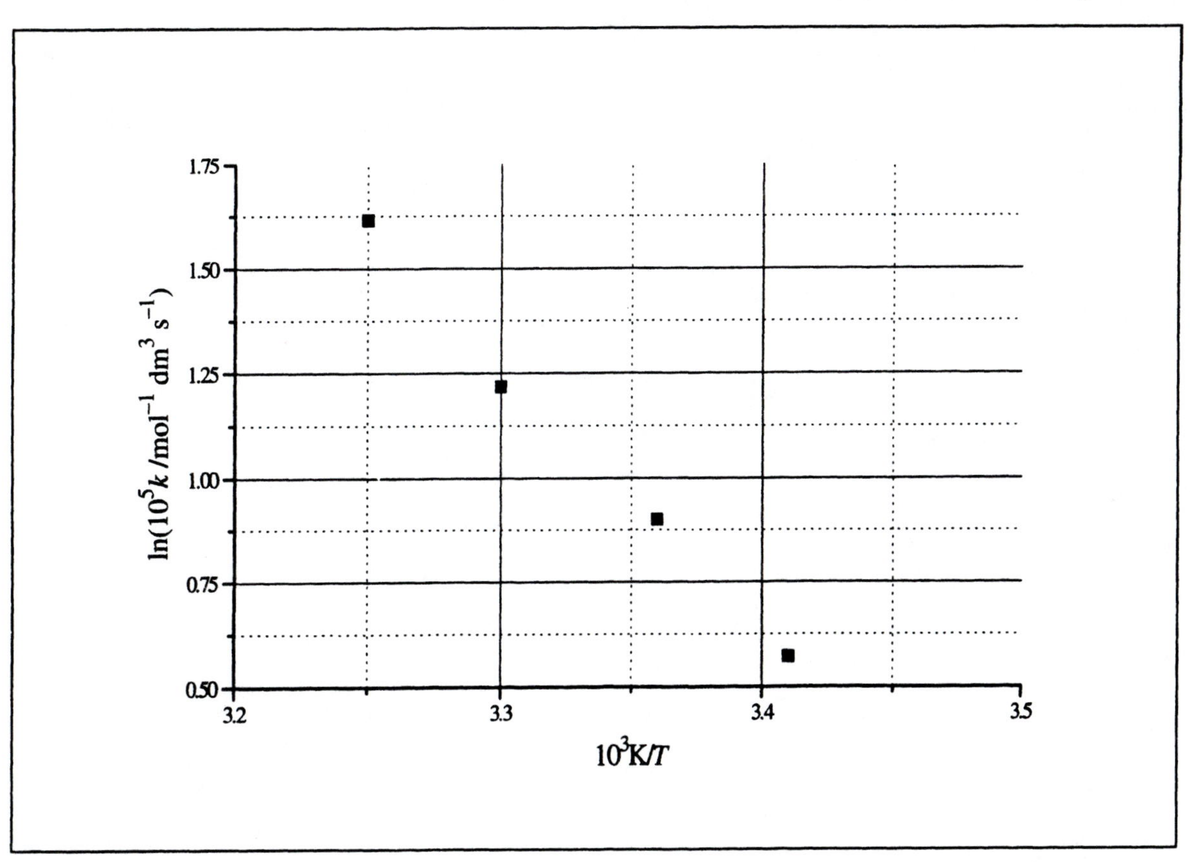
1.75
1.50
1.25
1.00
0.75
0.50
3.2
3.3
3.4
3.5
$\ln(10^5 k\ /\text{mol}^{-1}\ \text{dm}^3\ \text{s}^{-1})$
$10^3\text{K}/T$

SECTION 4

Calculating Frequencies, Wavelengths and Energies

Calculating Frequencies, Wavelengths and Energies

4.1 Converting between frequency, wavelength and wavenumber

The basic formulae and definitions required in this section are:

velocity = frequency × wavelength	$c = \nu \lambda$
and	
wavenumber = 1/wavelength	$\bar{\nu} = 1/\lambda$

From these we also get

frequency = velocity × wavenumber	$\nu = c\bar{\nu}$

Velocity has units of m s^{-1}.

The velocity of electromagnetic radiation is independent of frequency or wavelength: $c = 2.998 \times 10^8$ m s^{-1}.

Frequency has units of s^{-1} or Hz (these are the same thing).

Wavelength has units of m.

Wavenumber has units of m^{-1} (although the units cm^{-1} are commonly used, see below).

> High frequencies correspond to short wavelengths and high wavenumbers.
>
> Long wavelengths correspond to low frequencies and low wavenumbers.

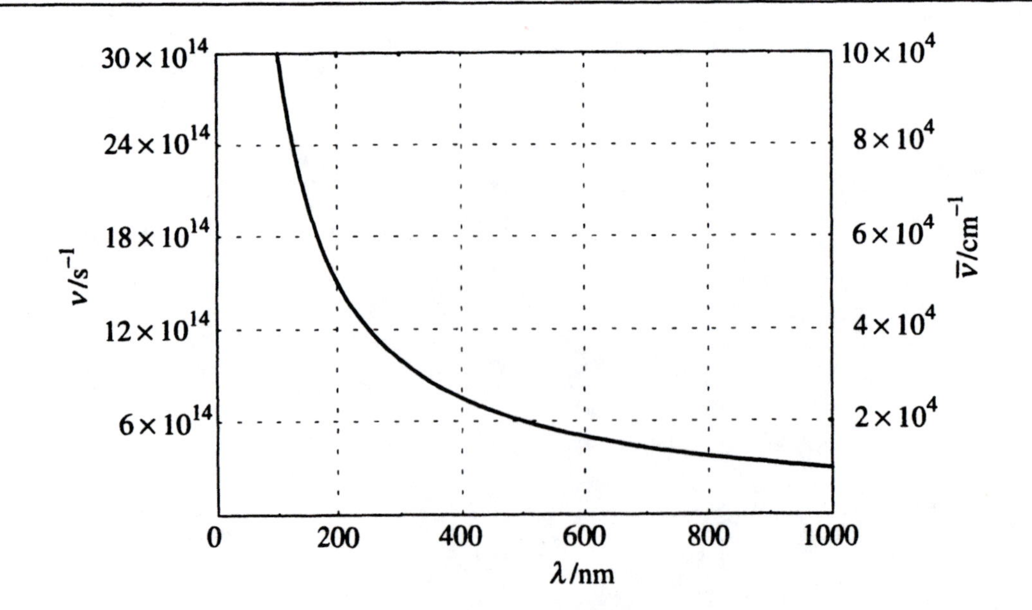

Example 4.1

Calculation of wavelength from frequency

Calculate the wavelength and wavenumber of light with frequency $6.4 \times 10^{14}\ \mathrm{s}^{-1}$

$$\lambda = \frac{c}{\nu} = \frac{2.998 \times 10^{8}\ \mathrm{m\,s}^{-1}}{6.4 \times 10^{14}\ \mathrm{s}^{-1}} = 4.68 \times 10^{-7}\ \mathrm{m} \qquad \bar{\nu} = \frac{1}{\lambda} = 2.13 \times 10^{6}\ \mathrm{m}^{-1}$$

This is light in the blue region of the visible spectrum.

Exercise 4.1

The frequencies of light corresponding to the other colours of the rainbow are as follows:

red, $\nu = 4.3 \times 10^{14}\ \mathrm{s}^{-1}$, orange, $\nu = 4.8 \times 10^{14}\ \mathrm{s}^{-1}$, yellow, $\nu = 5.2 \times 10^{14}\ \mathrm{s}^{-1}$,

green, $\nu = 5.7 \times 10^{14}\ \mathrm{s}^{-1}$, indigo, $\nu = 6.8 \times 10^{14}\mathrm{s}^{-1}$, violet, $\nu = 7.1 \times 10^{14}\mathrm{s}^{-1}$.

Calculate the corresponding wavelength and wavenumber in each case.

red: $\lambda = \frac{c}{\nu} = \rule{2cm}{0.4pt} =$ $\qquad \bar{\nu} = \frac{1}{\lambda} =$

orange: $\lambda = \frac{c}{\nu} = \rule{2cm}{0.4pt} =$ $\qquad \bar{\nu} = \frac{1}{\lambda} =$

yellow: $\lambda = \frac{c}{\nu} = \rule{2cm}{0.4pt} =$ $\qquad \bar{\nu} = \frac{1}{\lambda} =$

green: $\lambda = \frac{c}{\nu} = \rule{2cm}{0.4pt} =$ $\qquad \bar{\nu} = \frac{1}{\lambda} =$

indigo: $\lambda = \frac{c}{\nu} = \rule{2cm}{0.4pt} =$ $\qquad \bar{\nu} = \frac{1}{\lambda} =$

violet: $\lambda = \frac{c}{\nu} = \rule{2cm}{0.4pt} =$ $\qquad \bar{\nu} = \frac{1}{\lambda} =$

(Answers: $\lambda / \mathrm{nm} = 700,\ 620,\ 580,\ 530,\ 440,\ 420$; $\bar{\nu} / 10^{6}\mathrm{m}^{-1} = 1.43,\ 1.61,\ 1.72,\ 1.89,\ 2.27,\ 2.38$)

In many cases, wavenumbers are quoted in units of cm^{-1}. When using wavenumbers in the above equations for calculations, it is first necessary to convert from cm^{-1} to m^{-1}. This is achieved simply by multiplying by 100.

$$\bar{\nu} / \mathrm{m}^{-1} = 100 \times (\bar{\nu} / \mathrm{cm}^{-1})$$

The wavenumber for blue light corresponds to $\bar{\nu} = 21\,300\ \mathrm{cm}^{-1}$.

Example 4.2

Calculate the wavelength and frequency of radiation with wavenumber 2000 cm^{-1}

First, we convert the wavenumber: $\bar{\nu} = 2000\,cm^{-1} = 200\,000\,m^{-1}$ or $2 \times 10^5\,m^{-1}$. So

$$\lambda = \frac{1}{\bar{\nu}} = \frac{1}{2\times10^5\,m^{-1}} = 5\times10^{-6}\,m$$

$$\nu = c\bar{\nu} = 2.998\times10^8\,m\,s^{-1} \times 2\times10^5\,m^{-1} = 6\times10^{13}\,s^{-1}$$

This corresponds to radiation in the infra-red region of the spectrum.

Exercise 4.2

Calculate the wavelength and frequency of radiation from the microwave region with wavenumber $\bar{\nu} = 10\,cm^{-1}$

$$\bar{\nu} = 10\,cm^{-1} = \qquad m^{-1}$$

$$\lambda = \frac{1}{\bar{\nu}} = \frac{\qquad}{\qquad} =$$

$$\nu = c\bar{\nu} =$$

(Answers: $\bar{\nu} = 1\times10^3\,m^{-1}$, $\lambda = 1$ mm, $\nu = 3\times10^{11}\,s^{-1}$)

Exercise 4.3

(a) Calculate the wavelength and wavenumber of X-rays with frequency $3 \times 10^{17}\,s^{-1}$

$$\lambda = \frac{c}{\nu} = \frac{\qquad}{\qquad} = \qquad\qquad \bar{\nu} = \frac{1}{\lambda} =$$

(b) Calculate the frequency and wavenumber of UV radiation with wavelength 300 nm

$$\lambda = \frac{c}{\nu} = \frac{\qquad}{\qquad} = \qquad\qquad \bar{\nu} = \frac{1}{\lambda} =$$

(c) The radio-frequency region of the spectrum broadly corresponds to radiation with wavenumber less than 0.03 cm^{-1}. Calculate the wavelength and frequency at this wavenumber.

$\lambda = \frac{1}{\bar{\nu}} =$ cm = m $\quad\quad \nu = c\bar{\nu} =$

(Answers: (a) $\lambda = 1$ nm, $\bar{\nu} = 1\times10^{9}\,m^{-1}$; (b) $\nu = 1\times10^{15}\,s^{-1}$, $\bar{\nu} = 3.33\times10^{6}\,m^{-1}$; (c) $\lambda = 33$ cm = 0.33 m, $\nu = 9\times10^{8}\,s^{-1}$)

Exercise 4.4

Complete the following table for the frequencies and wavelengths of selected UK radio stations

station	waveband	frequency ν	wavelength λ
BBC Radio 1	FM MW MW	97.6–99.8 MHz 1053 kHz 1089 kHz	3.05–2.99 m 285 m 275 m
BBC Radio 2	FM	88.0–90.2 MHz	
BBC Radio 3	FM	90.2–92.4 MHz	
BBC Radio 4	FM LW	92.4–94.6 MHz	 1515 m
BBC Radio 5	MW MW	693 kHz	 328 m
BBC Radio Leeds	FM		3.23 m
BBC World Service	SW SW	6.195 MHz 9.410 MHz	
Classic FM	FM	100–102 MHz	
Virgin	MW	1197 kHz 1215 kHz	

Calculate typical wavenumber values for the frequency modulated (FM), long wave (LW), medium wave (MW) and short wave (SW) bands respectively.

(Answers, taking typical values from each of the ranges: FM, $\bar{\nu} = 0.33\,m^{-1}$; LW, $\bar{\nu} = 6.67\times10^{-4}\,m^{-1}$; MW, $\bar{\nu} = 3.6\times10^{-3}\,m^{-1}$; SW, $\bar{\nu} = 0.02\,m^{-1}$)

4.2 Energy, frequency and wavelength

The basic formula for this section relates the energy ε and frequency ν of radiation

$$\varepsilon = h\nu$$

where $h = 6.626 \times 10^{-34}$ J s is Planck's constant.

We may use the definitions given earlier to relate ε and λ:

$$\varepsilon = h\nu = \frac{hc}{\lambda} = hc\bar{\nu}$$

Here ε is the energy of a single photon and has units of J.

The corresponding 'molar energy' E, for a mole of photons all of the same frequency, is then obtained by introducing the Avogadro constant:

$$E = N_A h\nu = \frac{N_A hc}{\lambda} = N_A h\bar{\nu}$$

where we must include the units $\mathrm{mol^{-1}}$ with N_A.

Example 4.3

The frequency of blue light is $6.4 \times 10^{14}\ \mathrm{s^{-1}}$. Calculate the corresponding energy of (a) a single photon and (b) a mole of photons of this frequency.

(a) $\varepsilon = h\nu = 6.626 \times 10^{-34}\,\mathrm{Js} \times 6.4 \times 10^{14}\,\mathrm{s^{-1}} = 4.2 \times 10^{-19}\,\mathrm{J}$

(b) $E = N_A h\nu = 6.022 \times 10^{23}\,\mathrm{mol^{-1}} \times 6.626 \times 10^{-34}\,\mathrm{Js} \times 6.4 \times 10^{14}\,\mathrm{s^{-1}} = 255\,\mathrm{kJ\,mol^{-1}}$

Exercise 4.5

Repeat the above calculations for the remaining colours of the visible spectrum. Use your results to complete the table below.

red	$\varepsilon = h\nu =$	$E = N_A\varepsilon =$
orange	$\varepsilon = h\nu =$	$E = N_A\varepsilon =$
yellow	$\varepsilon = h\nu =$	$E = N_A\varepsilon =$
green	$\varepsilon = h\nu =$	$E = N_A\varepsilon =$
indigo	$\varepsilon = h\nu =$	$E = N_A\varepsilon =$
violet	$\varepsilon = h\nu =$	$E = N_A\varepsilon =$

	λ / nm	$\nu / 10^{14}\,s^{-1}$	$\varepsilon / 10^{-19}$ J	E / kJ mol^{-1}
red	700	4.3		
orange	620	4.8		
yellow	580	5.2		
green	530	5.7		
blue	470	6.4	4.2	255
indigo	440	6.8		
violet	420	7.1		

Answers (approximately): red $\varepsilon = 2.8 \times 10^{-19}$ J, E = 170 kJ mol^{-1}; orange $\varepsilon = 3.2 \times 10^{-19}$ J, E = 190 kJ mol^{-1}; yellow $\varepsilon = 3.4 \times 10^{-19}$ J, E = 205 kJ mol^{-1}; green $\varepsilon = 3.8 \times 10^{-19}$ J, E = 230 kJ mol^{-1}; indigo $\varepsilon = 4.5 \times 10^{-19}$ J, E = 270 kJ mol^{-1}; violet $\varepsilon = 4.7 \times 10^{-19}$ J, E = 285 kJ mol^{-1}.

4.3 Energy, frequency and wavelength of regions of the electromagnetic spectrum

Visible light typically corresponds to a wavelength of the order of 400–700 nm and to energies in the range 3–5 × 10^{-19} J or 170–290 kJ mol^{-1}.

Radiation from other regions of the spectrum corresponds to different ranges of wavelength and energy

Example 4.4

Calculate the energy of (i) one photon and (ii) a mole of photons of infra-red radiation of wavelength $\lambda = 5\times 10^{-6}$ m (i.e. for $\bar{\nu} = 2000\,\text{cm}^{-1}$).

(i) $$\varepsilon = \frac{hc}{\lambda} = \frac{6.626\times10^{-34}\,\text{Js}\times3\times10^{8}\,\text{m}\,\text{s}^{-1}}{5\times10^{-6}\,\text{m}} = 4\times10^{-20}\,\text{J}$$

(ii) $$E = N_A\varepsilon = 24\,\text{kJ}\,\text{mol}^{-1}$$

Approximate frequency and wavelength ranges for the different regions of the spectrum are:

X-ray and γ-rays:	$\lambda <$ 3 nm,	$\nu > 10^{17}\,s^{-1}$
Ultra-violet:	3 nm $< \lambda <$ 300 nm,	$10^{17}\,s^{-1} > \nu > 10^{15}\,s^{-1}$
Visible:	400 nm $< \lambda <$ 700 nm,	$10^{15}\,s^{-1} > \nu > 4\times10^{14}\,s^{-1}$
Infra-red:	1 μm $< \lambda <$ 3 mm,	$3\times10^{14}\,s^{-1} > \nu > 10^{11}\,s^{-1}$
Microwave:	3 mm $< \lambda <$ 30 cm,	$10^{11}\,s^{-1} > \nu > 10^{9}\,s^{-1}$
Radio-waves:	30 cm $< \lambda$	$10^{9}\,s^{-1} > \nu$

Exercise 4.6

Complete the following table, indicating the typical frequency, wavelength, wavenumber and energy of radiation from the different regions of the spectrum

	ν / s^{-1}	λ / m	$\bar{\nu} / cm^{-1}$	$\varepsilon / 10^{-19} J$	$E / kJ\ mol^{-1}$
X-rays	10^{17}	3×10^{-9}	3.3×10^{6}		
far UV	1.5×10^{15}	200×10^{-9}	50000		
near UV	1×10^{15}	300×10^{-9}	33000		
visible	6×10^{14}	500×10^{-9}	20000		
infra-red	3×10^{14}	1×10^{-6}	10000		
microwave	3×10^{11}	1×10^{-3}	10		
radio-waves	10^{9}	0.3	0.033		

Answers: $\varepsilon / 10^{-19}$ J = 660, 10, 6.6, 4, 2, 2×10^{-3} and 6.6×10^{-6}; *E*/kJ mol^{-1} = 4×10^{4}, 600, 400, 240, 120, 0.12 and 4×10^{-4}

Typical values

X-rays have energies in excess of 40 MJ mol^{-1}.

In the UV region, the energies range from 400 kJ mol^{-1} to 1 MJ mol^{-1}, corresponding to the typical strengths of chemical bonds

The visible region corresponds to the electronic transitions of some conjugated systems (e.g. dyes), with E ~240 kJ mol^{-1}.

The infra-red region corresponds to energies of the order of 100 kJ mol^{-1} which is typical of the separation between vibrational levels.

Microwaves have energies similar to the rotational energies of molecules, 100 J mol^{-1}.

The various regions of the electromagnetic spectrum and their associated frequency, wavelength and energy ranges are indicated in Figure 4.1 on the next page. You may mark on the figure the energies and molar energies of both the whole spectrum and the range corresponding to visible light which you have calculated. This figure may then be useful to you in the future.

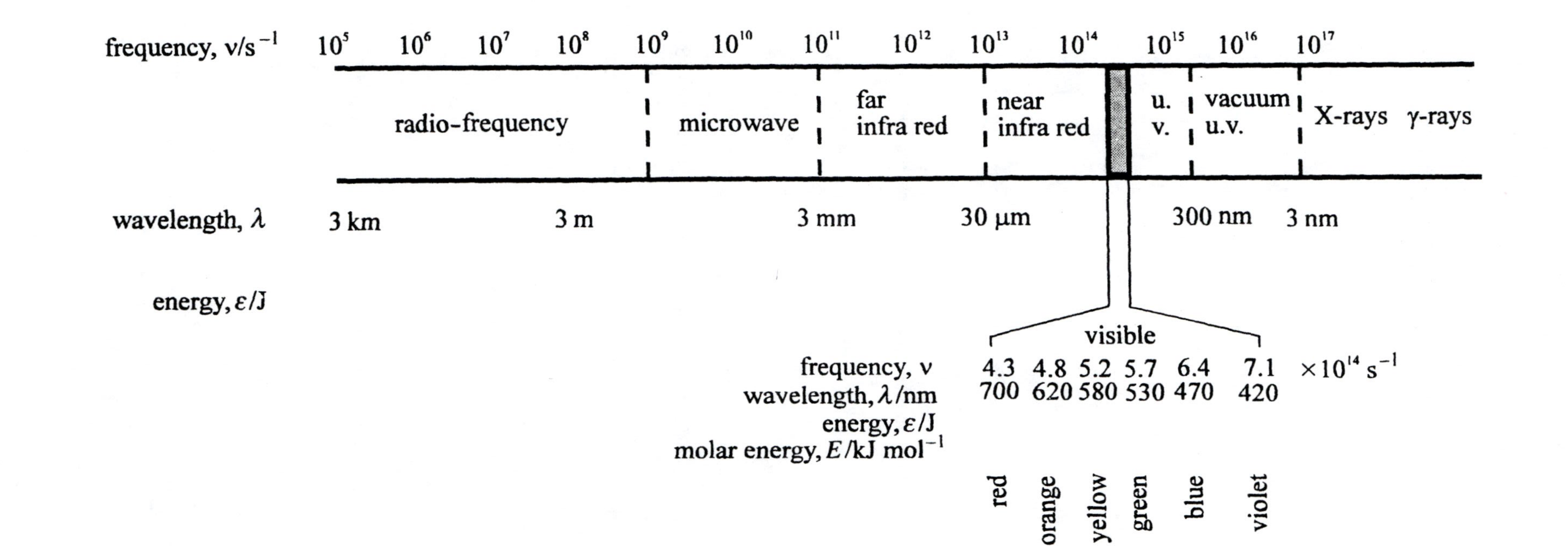

molar energy, E/kJ mol⁻¹
frequency, ν/s⁻¹
10⁵ 10⁶ 10⁷ 10⁸ 10⁹ 10¹⁰ 10¹¹ 10¹² 10¹³ 10¹⁴ 10¹⁵ 10¹⁶ 10¹⁷
radio-frequency
microwave
far infra red
near infra red
u. v.
vacuum u.v.
X-rays
γ-rays
wavelength, λ
3 km
3 m
3 mm
30 μm
300 nm
3 nm
energy, ε/J
visible
frequency, ν
4.3 4.8 5.2 5.7 6.4 7.1 ×10¹⁴ s⁻¹
wavelength, λ/nm
700 620 580 530 470 420
energy, ε/J
molar energy, E/kJ mol⁻¹
red
orange
yellow
green
blue
violet

Figure 4.1

4.4 Power

Much of our street lighting relies on the emission of light by *electronically excited* sodium atoms — the so-called sodium D-line emission that occurs at a wavelength $\lambda = 589\,\text{nm}$.

Exercise 4.7

What colour is the sodium D-line emission?

(Answer: yellow; the same emission is responsible for the intense yellow of the sodium flame test)

If we know how many photons per second a lamp is emitting and if we calculate the energy per photon, we can then determine the rate at which energy is being emitted, i.e. the *power* of the lamp at this wavelength.

Example 4.5

Determine the power of a sodium lamp that emits 2.965×10^{20} photons per second

energy of one photon $\quad \varepsilon = \dfrac{hc}{\lambda} = 3.37 \times 10^{-19}\,\text{J}$

power of lamp at D-line $\quad P = 2.965 \times 10^{20}\,\text{s}^{-1} \times 3.37 \times 10^{-19}\,\text{J} = 100\,\text{W}$

Note: it is conventional not to include 'photon' in the units, so the energy is 3.37×10^{-19} J not 3.37×10^{-19} J photon^{-1} and the flux is $2.965 \times 10^{20}\,\text{s}^{-1}$ not 2.965×10^{20} photon s^{-1}

Exercise 4.8

(a) Calculate the power of a neon lamp emitting 1.5×10^{20} photons per second at $\lambda = 632.8\,\text{nm}$

(b) How many photons are emitted per second by an Hg lamp operating at 500 W at a wavelength $\lambda = 254\,\text{nm}$?

(c) The continuous-wave (CW) CO_2 laser is used to precision-cut metals, operating with an emission of wavelength $\lambda = 10.6\,\mu\text{m}$ and at a power of up to 10 kW. Calculate the rate of photon emission under these conditions.

(Answers: (a) 47.1 W; (b) $6.4 \times 10^{20}\,\text{s}^{-1}$; (c) $5.34 \times 10^{23}\,\text{s}^{-1}$, i.e. 0.886 mol s^{-1})

Returning to the sodium D-line emission; if this is inspected with a high-resolution spectrophotometer, it is actually seen to consist of *two* closely spaced lines of wavelengths $\lambda_1 = 589.2$ nm and $\lambda_2 = 589.8$ nm.

Exercise 4.9

Calculate the splitting of these two lines in terms of the difference in their wavenumbers and the difference in their energy.

(Answers: $\Delta\bar{\nu} = \dfrac{1}{589.2\times10^{-9}\,\text{m}} - \dfrac{1}{589.8\times10^{-9}\,\text{m}} = 1727\,\text{m}^{-1} = 17.3\,\text{cm}^{-1}$; $\Delta\varepsilon = 3.43 \times 10^{-22}$ J)

4.5 The photoelectric effect

If a surface is illuminated by light of a suitable wavelength, then electrons are emitted by the atoms in the surface. This is known as the *photoelectric effect.*

It is observed that electrons are only emitted if the wavelength of the illuminating photons is below some critical threshold. This is rationalised by noting that, as the wavelength decreases, so the energy of the photons increases. Thus, there is a minimum energy below which emission of photoelectrons does not occur.

This minimum energy is known as the *work function* (or, also, as the binding energy of the electron or the ionisation energy). The value of the work functions is characteristic of the particular atom from which the electron is being ejected, and this forms the basis of an analytical technique known as *photoelectron spectroscopy.*

If the wavelength is less than the critical wavelength, then the electrons emitted can be detected and their kinetic energy measured. The relationship between the kinetic energy ε_k, the work function ϕ and the wavelength λ of the incident photon is

$$\varepsilon_k = \frac{hc}{\lambda} - \phi$$

Exercise 4.10

Explain the form of this equation

(Possible answer: the first term on the right-hand side is the energy of the incoming photon, the right-hand side thus gives the difference between the energy coming in and the energy required to eject the electron from its bound energy level; this 'excess' energy is carried away as kinetic energy in the emitted electron.)

The 'critical' case is that at which the electron is emitted with zero kinetic energy, i.e. for which $\varepsilon_k = 0$ in the above equation. This gives the maximum wavelength above which the photoelectric effect does not arise for a given atom.

Exercise 4.11

Calculate the wavelength below which photoelectrons are expected to be emitted from potassium atoms if the work function $\phi_K = 3.56 \times 10^{-19}$ J

(Answer: for $\varepsilon_k = 0$, $\lambda = hc / \phi = 558$ nm (yellow/green light))

The kinetic energy of an emitted electron will be related to its velocity v by $\varepsilon_k = \frac{1}{2} m_e v^2$ where m_e is the mass of the electron.

Exercise 4.12

Calculate the expected kinetic energy and velocity of a photoelectron emitted by a potassium surface subject to light of wavelength $\lambda = 530\,\text{nm}$

We can note that the wavelength is less than the critical wavelength calculated in the previous exercise, so we do indeed expect photoelectrons to be emitted with some positive kinetic energy:

$\varepsilon_k =$

so then we can find v

$v =$

(Answers: $\varepsilon_k = 1.88 \times 10^{-20}$ J, $v = 451$ m s^{-1})

4.6 Other units of energy

So far, we have expressed all our energy values in the SI units of J or J mol^{-1}. Some other units are fairly common in chemistry and it is important to be able to convert from these into the SI values for calculations.

The *calorie* is an old unit of energy, originally defined in terms of the energy required to raise the temperature of 1 g of water by 1 kelvin under specified conditions of temperature and pressure. Now, however, the calorie (symbol cal) is defined explicitly in terms of the joule:

$$1 \text{ cal} = 4.184 \text{ J}$$

To convert an energy *from* calories *to* joules, therefore, we *multiply* the numerical value by 4.184 (the joule is the smaller unit, therefore an energy is a larger number of joules than calories).

To convert an energy *to* calories *from* joules, we *divide* the numerical value by 4.184

Example 4.6

The bond dissociation energy D in the H_2 molecule is 436 kJ mol^{-1}. We can express this in calories using the above rules

$$D(H_2) = 436\,\text{kJ}\,\text{mol}^{-1} = \frac{436}{4.184}\,\text{kcal}\,\text{mol}^{-1} = 104.2\,\text{kcal}\,\text{mol}^{-1}$$

The activation energy for the reaction between an H atom and an O_2 molecule has the value E_A = 16.5 kcal mol^{-1}. The value of the activation energy in SI units is thus

$$E_A = 16.5 \times 4.184\,\text{kJ}\,\text{mol}^{-1} = 69.0\,\text{kJ}\,\text{mol}^{-1}$$

The Calorie, with a capital initial letter, is used in quoting the energy value of foods on packaging and in dietary advice: 1 Cal = 1 kcal = 4.184 kJ.

Exercise 4.13

(a) The energy ε of a photon of blue light is 4.2×10^{-19} J. Express this in calories.

(b) The specific heat capacity of liquid water at 298 K has the value C = 1.00 cal K^{-1} g^{-1}. Convert this to SI units.

(c) A 100 g serving of a particular breakfast cereal provides 370 Cal. What is this energy (i) in J and (ii) in J kg^{-1}?

(Answers: (a) 1.00×10^{-19} cal; (b) C = 4.184 kJ K^{-1} kg^{-1}; (c) (i) 1.55 MJ, (ii) 15.5 MJ kg^{-1}

The second unit of interest is the *electron volt* (symbol eV). This is employed most often when we wish to represent large energy values.

The definition of the electron volt is the energy gained by an electron as it is accelerated by a potential of 1 volt. The value of 1 eV can be calculated from this definition by multiplying the charge on an electron by 1 V, to give

$$1\,\text{eV} = 1.602177 \times 10^{-19}\,\text{C} \times 1\,\text{V} = 1.602177 \times 10^{-19}\,\text{J}$$

(Note that charge × voltage = energy, so coulomb × volt = joule)

This may not seem a particularly large energy value, but we can appreciate its magnitude more by multiplying by the Avogadro constant to obtain the equivalent molar energy

$$1\,\text{eV} \equiv 96.485\,\text{kJ}\,\text{mol}^{-1}$$

so, as a rule of thumb for quick estimates we can take 1 eV ≈ 100 kJ mol^{-1}.

To convert an energy expressed in J to one in eV, we divide the numerical value by 1.602×10^{-19}

To convert an energy to J from one expressed in eV, we multiply the numerical value by 1.602×10^{-19}

To convert an energy expressed in kJ mol^{-1} to one in eV, we divide the numerical value by 96.485

To convert an energy to kJ mol^{-1} from one expressed in eV, we multiply the numerical value by 96.485

Example 4.7

The energy of blue light is 4.2×10^{-19} J. What is this in eV?

$$\varepsilon = 4.2 \times 10^{-19}\,\text{J} = \frac{4.2 \times 10^{-19}}{1.602 \times 10^{-19}}\,\text{eV} = 2.6\,\text{eV}$$

It is also quite common in older texts and even in modern research papers to find spectroscopists quoting 'energies' in terms of '*wavenumbers*'. To convert from these convenient units to the SI forms needed in calculations, we can use the equation $\varepsilon = hc\overline{\nu}$.

Thus, the energy equivalent to 'one wavenumber', i.e. to $\overline{\nu} = 1\,\text{cm}^{-1}$ can be calculated. Remembering that 1 cm^{-1} = 100 m^{-1}, we have $\varepsilon = 1.986 \times 10^{-23}$ J

i.e. $1\ \text{cm}^{-1} \equiv 1.986 \times 10^{-23}\ \text{J} \equiv 11.96\ \text{J mol}^{-1}$

The second form, giving the equivalent molar energy, comes from multiplying by N_A.

Exercise 4.14

(a) Calculate the energy of a 0.4 keV X-ray

(b) What is the actual energy of a spectroscopic transition reported as corresponding to 25 cm^{-1}?

(c) Work functions are usually quoted in electron volts: express the value of $\phi_K = 3.56 \times 10^{-19}$ J in these units

(Answers: (a) 0.4 keV = 400 eV = 400 × 96.485 kJ mol^{-1} = 38.6 MJ mol^{-1}; (b) 25 × 1.986×10^{-23} J = 4.97×10^{-22} J or 300 J mol^{-1}; (c) 2.22 eV)

Important Equations used in this Section

The following should be familiar after completing this section and are gathered here for reference:

relationship between velocity, frequency and wavelength

$$c = \nu \lambda$$

definition of wavenumber $\quad \bar{\nu} = 1/\lambda = \nu/c$

conversion factor $\quad \left(\bar{\nu}/\mathrm{m}^{-1}\right) = 100 \times \left(\bar{\nu}/\mathrm{cm}^{-1}\right)$

Planck formula $\quad \varepsilon = h\nu = \dfrac{hc}{\lambda} = hc\bar{\nu}$

Molar energy $\quad E = N_A\varepsilon = N_A h\nu = \dfrac{N_A hc}{\lambda} = N_A hc\bar{\nu}$

Conversion factors

$1\ \mathrm{cal} = 4.184\ \mathrm{J}$
$1\ \mathrm{eV} = 1.602 \times 10^{-19}\ \mathrm{J} \equiv 96.485\ \mathrm{kJ\ mol^{-1}}$
$1\ \mathrm{cm^{-1}} = 1.986 \times 10^{-23}\ \mathrm{J} \equiv 11.96\ \mathrm{J\ mol^{-1}}$.

SECTION 5

Pressure, Volume, Temperature: Concentration and Density

Pressure, Volume, Temperature: Concentration and Density

5.1 The equation of state for a perfect gas

The pressure p, volume V and temperature T of a *perfect* or *ideal gas* are related by the following *equation of state*:

$$pV = nRT$$

Here, the quantity n is called the *amount of substance* and has units of moles (symbol mol)

The SI unit for pressure, p, is the pascal (symbol Pa) which is equivalent to newton per square metre N m^{-2}, (i.e. force per unit area) and is also the same as J m^{-3} (i.e. energy per unit volume).

The SI unit for volume, V, is m^3

The SI unit for temperature, T, is kelvin, K

The Gas constant R has the value and units: $R = 8.314\ \mathrm{J\,K^{-1}\,mol^{-1}}$

Other units of measure for p and V will be discussed later.

Example 5.1

Calculate the pressure exerted by one mole of He in a volume of 1×10^{-3} m^3 at 300 K assuming the gas behaves ideally.

$$p = \frac{nRT}{V} = \frac{1\,\mathrm{mol} \times 8.314\,\mathrm{J\,K^{-1}\,mol^{-1}} \times 300\,\mathrm{K}}{1 \times 10^{-3}\,\mathrm{m^3}} = 2.49 \times 10^6\ \mathrm{J\,m^{-3}} = 2.49 \times 10^6\,\mathrm{Pa}$$

Exercise 5.1

Calculate the volume occupied by 2 mol of He at a pressure of 1×10^5 Pa and at 500 K assuming the gas behaves ideally.

$V =$ ———— $=$ ———————————— $=$

(Answer, $V = 83 \times 10^{-3}$ m^3: noting that the units of J/Pa = m^3)

5.2 Standard pressure

The standard pressure p^θ is now defined as being a pressure of exactly 1×10^5 Pa

$$p^\theta = 1 \times 10^5 \text{ Pa}$$

Example 5.2

Express the pressure from the previous worked example in terms of p^θ

In the previous example, we calculated $p = 2.49 \times 10^6$ Pa

$$\frac{p}{p^\theta} = \frac{2.49 \times 10^6 \text{ Pa}}{1 \times 10^5 \text{ Pa}} = 24.9, \qquad \text{i.e. } p = 24.9\, p^\theta$$

Exercise 5.2

The pressure at a depth of 1500 m of water is approximately 10×10^6 Pa. Express this in terms of the standard pressure.

$$\frac{p}{p^\theta} = \underline{\qquad\qquad\qquad} = \qquad\qquad \text{i.e. } p =$$

(Answer: $p = 100\, p^\theta$)

5.3 Molar volume

Colloquially, the molar volume V_m is simply the volume occupied by one mole of gas.

More precisely, the molar volume is the volume per unit amount of substance

$$V_m = V/n$$

For a perfect gas

$$V_m = \frac{V}{n} = \frac{RT}{p}$$

Example 5.3

Calculate the standard molar volume V_m^θ of a perfect gas at 298 K.

The standard molar volume is a special case of V_m: it is the molar volume corresponding to a particular pressure: $p = p^\theta = 10^5$ Pa. So

$$V_m^\theta = \frac{8.314\,\text{J K}^{-1}\,\text{mol}^{-1} \times 298\,\text{K}}{1 \times 10^5\,\text{Pa}} = 24.78 \times 10^{-3}\,\text{m}^3\,\text{mol}^{-1}$$

remembering that Pa = J m^{-3}.

Molar volumes have units of volume per amount of substance: $m^3\ mol^{-1}$.

Molar volumes can be calculated at any pressure and temperature.
Standard molar volumes are evaluated with $p = p^\theta$ and can be calculated for any temperature.

Exercise 5.3

(a) Calculate the molar volume of a perfect gas for $p = 2.5 \times 10^4$ Pa at $T = 298$ K

$$V_m = \frac{RT}{p} = \underline{\hspace{6cm}} =$$

(b) Calculate the molar volume of a perfect gas for $p = p^\theta$ at T = 5000 K

(Answers: (a) $V_m = 0.1\ m^3\ mol^{-1}$; (b) $V_m = 0.42\ m^3\ mol^{-1}$)

5.4 Boyle's law: volume and pressure at constant temperature and amount of substance

Robert Boyle's name has become associated with the observation that the volume occupied by a gas is inversely proportional to the applied pressure if n and T are held constant.

This statement can be expressed in a number of ways mathematically:

$$V \propto \frac{1}{p} \qquad \text{or} \qquad V = \frac{\text{const}}{p} \qquad \text{or} \qquad pV = \text{const}$$

where const indicates a constant.

For a perfect gas, const = nRT.

The *p–V isotherms* corresponding to this relationship for a perfect gas at various temperatures are sketched in the figure below:

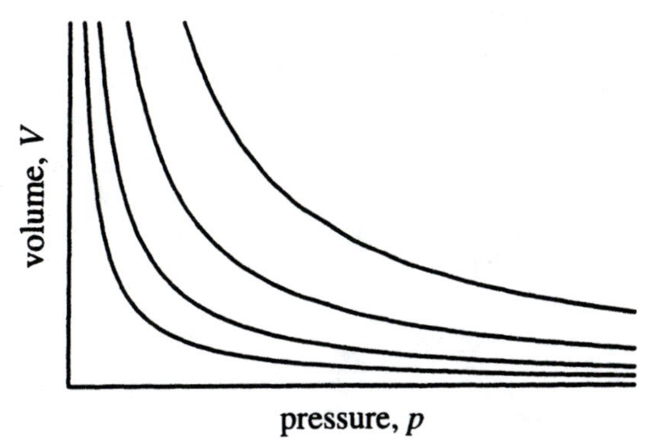

Doubling the pressure halves the volume: halving the pressure doubles the volume.

Exercise 5.4

Which curves in the figure correspond to the highest temperatures?

(Answer: increasing the temperature increases the value of the constant, so the higher curves correspond to the larger values of T)

Boyle's law can be used to find the change in volume during an isothermal (at constant temperature) expansion or compression of a perfect gas.

Example 5.4

A sample of gas has an initial volume of 24.78×10^{-3} m^3 at a pressure of 1×10^5 Pa. What is the volume if the pressure is decreased to 1×10^4 Pa?

If we denote the initial pressure and volume by p_1 and V_1, we can use these to evaluate the constant:

$$p_1V_1 = \text{const} = 1\times10^5\,\text{Pa} \times 24.78\times10^{-3}\,\text{m}^3 = 2478\,\text{J}$$

The final pressure p_2 and volume V_2 must satisfy the same equation, *with the same value of the constant*, so

$$p_2V_2 = 2478\,\text{J} \qquad \text{so} \qquad V_2 = \frac{2478\,\text{J}}{10^4\,\text{Pa}} = 247.8\times10^{-3}\,\text{m}^3$$

The 10-fold decrease in the pressure results in a 10-fold increase in the volume

Rather than explicitly evaluating the constant in this calculation, we can note that since both p_1V_1 and p_2V_2 are equal to the same constant, they must also be equal to each other:

$p_1V_1 = \text{const} = p_2V_2$, so

$$p_1V_1 = p_2V_2 \qquad \text{or} \qquad \frac{p_1}{p_2} = \frac{V_2}{V_1}$$

Example 5.5

Calculate the final volume for the previous system if the final pressure is increased by a factor of ten to 1×10^6 Pa

Using the above relationships

$$V_2 = \frac{p_1}{p_2} V_1 = \frac{1 \times 10^5 \, \text{Pa}}{1 \times 10^6 \, \text{Pa}} \times 24.78 \times 10^{-3} \, \text{m}^3 = 2.478 \times 10^{-3} \, \text{m}^3$$

The volume decreases by a factor of 10

Another point of interest here is that we have to calculate the ratio of the two pressures p_1/p_2. The units of pressure thus cancel out in a clear way. This also means that if the pressures had been quoted in some other system of units, e.g. in 'millimetres of mercury' mmHg, then there would be no need to convert them to Pa before performing the calculation as the same conversion factor would be involved in both the top and bottom of this ratio.

Exercise 5.5

(a) A hot-air balloon ascends to a height at which the surrounding air pressure is one-fifth of that on the ground. Assuming that the temperature of the gas remains constant and that no gas escapes, what would be the change in the volume of the balloon?

Find an expression for V_2/V_1 in terms of $p_1$and p_2 and then substitute in from the given information:

$$\frac{V_2}{V_1} = \frac{\quad}{\quad} =$$

(b) A perfect gas initially at 0.5×10^5 Pa and occupying a volume of 1×10^{-3} m^3 expands to 5/3 of its initial volume. Calculate the new pressure.

$$p_2 =$$

(c) Calculate the pressure to which a perfect gas must be raised to cause a 6-fold decrease in volume if the initial pressure is 2×10^4 Pa

(d) A perfect gas at an initial pressure of 760 mmHg is expanded so that the pressure falls to 300 mmHg. If the initial volume was 0.05 m^3, what will the final volume be?

(Answers: (a) $V_2/V_1 = p_1/p_2 = 5$, the volume increases to five times its initial value; (b) $p_2 = (3/5)p_1 = 0.3 \times 10^5$ Pa; (c) $p_2 = 6p_1 = 1.2 \times 10^5$ Pa, (d) $V_2 = (760 \text{ mmHg}/300 \text{ mmHg}) \times 0.05 \text{ m}^3 = 0.127 \text{ m}^3$)

5.5 Charles's law: volume and temperature at constant pressure and amount of substance

Jacques Charles noted that the volume of a gas is proportional to the absolute temperature provided the pressure and amount of substance remain constant:

$$V \propto T \qquad \text{or} \qquad V = \text{const} \times T$$

For a perfect gas, the constant is given by const = nR/p.

Plots of V against T are thus straight lines with gradients that increase with the amount of substance but decrease with increasing pressure:

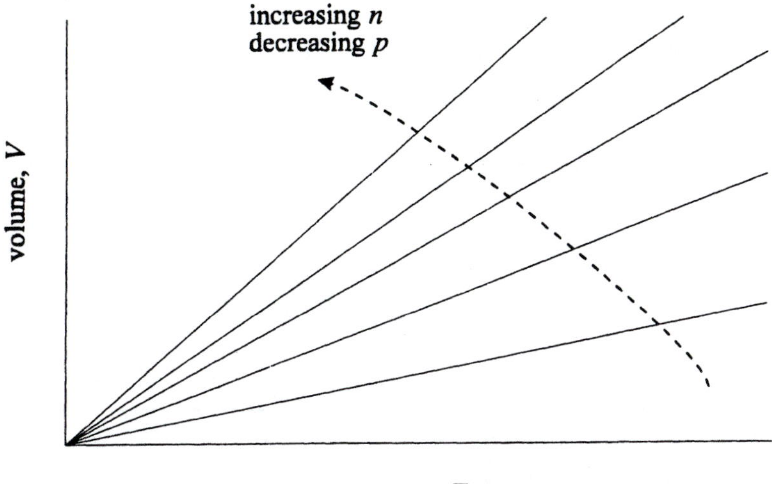

Doubling the temperature doubles the volume: halving the temperature halves the volume.

Charles's law can be re-written as

$$\frac{V_1}{V_2} = \frac{T_1}{T_2} \qquad \text{or} \qquad \frac{V_1}{T_1} = \frac{V_2}{T_2}$$

where V_1 and T_1 are the initial volume and temperature and V_2 and T_2 are the final volume and temperature respectively.

Exercise 5.6

(a) The molar volume V_m of a perfect gas at standard pressure and at 298 K is 24.78×10^{-3} m^3. Calculate the molar volume at (i) 500 K and (ii) 50 K

(i) $$V_{m,2} = \frac{T_2}{T_1} \times V_{m,1} =$$

(ii) $$V_{m,2} = \frac{\quad}{\quad}$$

(b) A balloon is filled with 0.2 mol of He at standard pressure and at 273 K. Calculate the volume of the balloon under these conditions and then the volume occupied if the balloon is cooled in liquid nitrogen to 77 K.

$$V_1 = \frac{nRT_1}{p} = \underline{\hspace{6cm}} =$$

and then

$$V_2 = \underline{\hspace{2cm}} \times \quad =$$

(Answers: (a) (i) 41.6×10^{-3} m^3; (ii) 4.16×10^{-3} m^3; (b) $V_1 = 4.54 \times 10^{-3}$ m^3, $V_2 = 1.28 \times 10^{-3}$ m^3

The equation of state for an ideal gas (Section 5.1) can be combined with the Boyle and Charles laws to produce a single equation which expresses the interdependency of the pressure, volume and temperature of a sample:

$$\frac{p_1 V_1}{T_1} = \frac{p_2 V_2}{T_2}$$

Exercise 5.7

The fully burned gases in the combustion chambers of a medium size family car reach a temperature of 2000 K and a pressure of 200 bar, where 1 bar $\equiv 1 \times 10^5$ Pa. What volume will these gases occupy after being exhausted in to the atmosphere on a still summer's day? Choose appropriate conditions for the weather and volume of gases in the car's engine and use the equation above.

(Answer: V should be somewhere between 27 and 96 litres depending on your own choices. A typical value for a 2 litre engine, and a temperature of 20°C and pressure of 1 bar would be 60 litres.)

5.6 Concentration and density

The concentration c of a substance is the amount of substance per unit volume

$$c = \frac{n}{V}$$

The SI unit of concentration is, therefore, mol m^{-3}.

A number of other units are frequently more convenient (such as mol dm^{-3} for solution-phase reactions): some of these will be discussed in the next section.

Exercise 5.8

Re-arrange the ideal gas equation $pV = nRT$ to give an expression for the concentration of a gas in terms of the pressure and temperature:

$$c = \frac{n}{V} = \frac{\quad}{\quad}$$

(Answer: $c = p/RT$)

Example 5.6

Calculate the concentration of an ideal gas at the standard pressure p^{θ} and at 298 K

$$c = \frac{p^{\theta}}{RT} = \frac{1 \times 10^5\,\text{Pa}}{8.314\,\text{J K}^{-1}\text{mol}^{-1} \times 298\,\text{K}} = 40.36\,\text{mol m}^{-3}$$

Notes

The inverse of this quantity, $1/c = 24.78 \times 10^{-3}$ m^3 mol^{-1}, is the standard molar volume calculated previously in Section 5.3.

The units emerge naturally: remembering that Pa is equivalent to J m^{-3} (see Section 3.3), the J in the numerator and denominator cancel, as do the K^{-1} from R and the K from the temperature in the denominator: mol^{-1} in the denominator is equivalent to mol in the numerator.

Exercise 5.9

Calculate the concentration of an ideal gas under the following conditions of pressure and temperature:

(a) $p = 0.5 \times 10^5$ Pa, $T = 298$ K

$$c = \frac{p}{RT} = \text{—————————} =$$

(b) $p = 1 \times 10^5$ Pa, $T = 398$ K

$c =$

(c) Compare your results in (a) and (b) above with that for the standard conditions of the worked example: what is the effect on the concentration of a gas of

(i) reducing the pressure

(ii) increasing the temperature

(d) Calculate the concentration of an ideal gas under the following conditions of pressure and temperature

(i) $p = 1 \times 10^6$ Pa, $T = 500$ K

$c =$

(ii) $p = 1 \times 10^4$ Pa, $T = 500$ K

$c =$

(Answers: c/mol m^{-3} = (a) 20.18; (b) 30.22; (c) (i) reducing the pressure decreases the concentration, (ii) increasing the temperature decreases the concentration; (d) (i) 240.6, (ii) 2.406)

The density ρ of a sample is the mass per unit volume

$$\rho = \frac{m}{V}$$

The SI unit of density, therefore, is kg m^{-3}.

The density and concentration of a pure substance are related via the molar mass M (see Section 2.3).

The mass m of a sample with amount of substance n and molar mass M is:

$$m = M n$$

The density is then given by

$$\rho = \frac{m}{V} = \frac{M n}{V} = M c$$

Note: the units of M (kg mol^{-1}) and of concentration (mol m^{-3}) multiply together to produce those of density (kg m^{-3})

Example 5.7

Calculate the density of nitrogen under standard conditions ($p = p^{\theta}$ and $T = 298$ K)

From the worked example of the previous section we have $c = 40.36$ mol m^{-3}

$M(N_2) = 28$ g $mol^{-1} = 28 \times 10^{-3}$ kg mol^{-1}

so $$\rho = Mc = 28 \times 10^{-3}\,\text{kg} \times 40.36\,\text{mol m}^{-3} = 1.13\,\text{kg m}^{-3}$$

This density is equivalent to 1.13 g dm^{-3}: a litre of air weighs approximately 1 g under standard conditions.

The mass of a sample of gas is given by

$$m = Mn = McV = M\frac{pV}{RT}$$

Exercise 5.10

(a) Substitute for the concentration $c = p/RT$ to obtain an expression for the density ρ in terms of the molar mass M, the pressure p and the temperature T

$\rho =$ ——

(b) Calculate the density of helium gas at the pressures and temperatures for the previous set of exercises:

(i) $p = 0.5 \times 10^5$ Pa, $T = 298$ K
(ii) $p = 1 \times 10^5$ Pa, $T = 398$ K
(iii) $p = 1 \times 10^6$ Pa, $T = 500$ K
(iv) $p = 1 \times 10^4$ Pa, $T = 500$ K

In each case, also calculate the mass of a sample of volume 24.78×10^{-3} m^3.

(i)

$\rho =$

$m =$

(ii)

(iii)

(iv)

(Answers: (a) $\rho = Mp/RT$ (b) ρ/kg m^{-3}, m/g: (i) 0.08, 1.98; (ii) 0.121 3.0; (iii) 0.962, 23.8; (iv) 9.62×10^{-3}, 0.238)

5.7 Other units for concentration, volume and pressure

The SI units of mol m^{-3}, m^3 and Pa for c, V and p respectively are not always the most convenient experimentally.

In solution-phase reactions, concentrations are more often expressed in terms of moles per litre (i.e. mol dm^{-3}), with the symbol M.

Atmospheric chemists frequently find it convenient to express concentrations in terms of molecules per cubic centimetre.

Pressures are often measured experimentally with a mercury manometer, so the pressures are frequently reported in the unit of 'millimetres of mercury', symbol mmHg or the related unit of Torr.

In the current engineering literature, the imperial unit of pounds per square inch, symbol lb in^{-2}, can still be found.

(i) converting between volume units

The SI unit of 1 cubic metre is generally too large for chemical relevance: we tend to use litre flasks etc.

The litre is defined as one cubic decimetre: 1 L = 1 dm^3

When performing calculations, however, it is generally best to convert through to the SI units before entering quantities in equations.

To convert from a volume V quoted or measured in litres to m^3 we use the following idea:

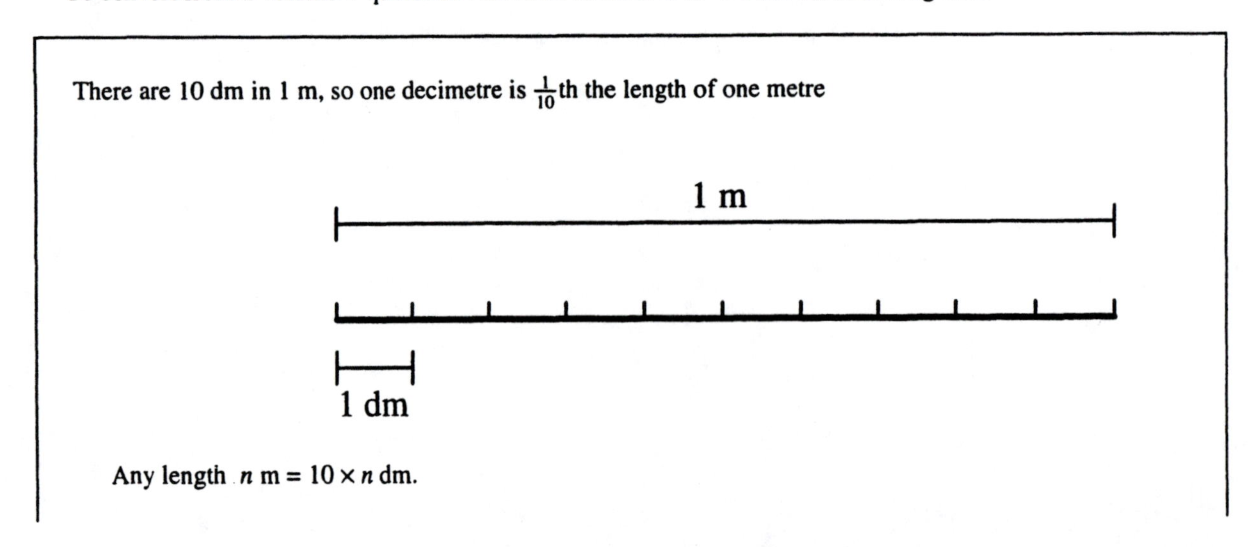

If we now consider areas, a square with sides of length 1 m will have a total area of 1 m^2 (by definition). This square will contain $10 \times 10 = 100$ squares of area 1 dm^2

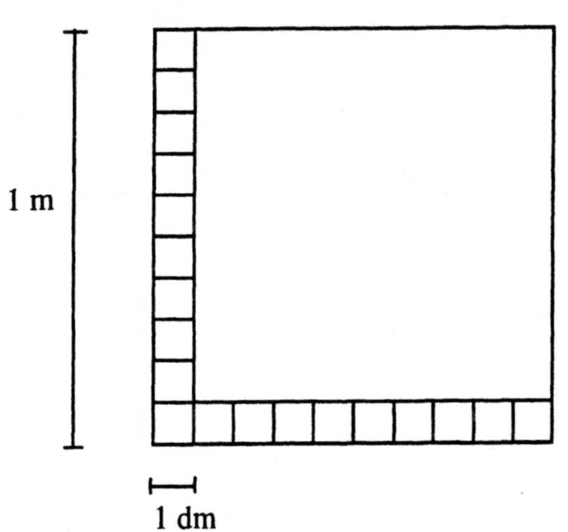

1 dm^2 is 100 times smaller than 1 m^2: any area will comprise 100 times *more* dm^2 than m^2

For a cube, there will be $10 \times 10 \times 10 = 1000$ small cubes of volume 1 dm^3 in a large cube of volume 1 m^3, so 1 dm^3 is 1000 times smaller than 1 m^3. A volume expressed in dm^3 will have a numerical factor 1000 times larger than the same volume expressed in m^3

More formally, the conversion factor can be derived using 'unit calculus'

By definition $\quad 1\ \text{dm} = 0.1\ \text{m}$

therefore, taking the cube for volume

$$(1\ \text{dm})^3 = (0.1\ \text{m})^3$$

i.e. $$1\ \text{dm}^3 = (0.1)^3\ \text{m}^3$$

and so $$1\ \text{L} = 1\ \text{dm}^3 = 10^{-3}\ \text{m}^3$$

Example 5.8

A chemical reactor has a volume of 3.7 L, what is the volume in cubic metres?

$$V = 3.7\ \text{L} = 3.7 \times (1 \times 10^{-3}\ \text{m}^3) = 3.7 \times 10^{-3}\ \text{m}^3$$

Exercise 5.11

Express the following in m^3

(a) 5 L

(b) 36.1 dm^3

(Answers: (a) $5 \times 10^{-3}\ \text{m}^3$; (b) $36.1 \times 10^{-3}\ \text{m}^3$)

The units of millilitre (1 mL = 1×10^{-3} dm^3) and cubic centimetre (cm^3) are equivalent:

$$1 \text{ mL} = 1 \text{ cm}^3 = (0.01 \text{ m})^3 = 10^{-6} \text{ m}^3$$

There are one million cubic centimetres in a cubic metre

Exercise 5.12

Find the relationship between mm^3 and m^3

(Answer: $1 \text{ mm}^3 = 10^{-9} \text{ m}^3$ or $10^9 \text{ mm}^3 = 1 \text{ m}^3$)

Conversion factors

	to m^3	from m^3
	$1 \text{ L} = 1 \times 10^{-3} \text{ m}^3$	$1 \text{ m}^3 = 10^3 \text{ L}$
	$1 \text{ dm}^3 = 1 \times 10^{-3} \text{ m}^3$	$1 \text{ m}^3 = 10^3 \text{ dm}^3$
	$1 \text{ cm}^3 = 1 \times 10^{-6} \text{ m}^3$	$1 \text{ m}^3 = 10^6 \text{ cm}^3$
	$1 \text{ mm}^3 = 1 \times 10^{-9} \text{ m}^3$	$1 \text{ m}^3 = 10^9 \text{ mm}^3$
also useful is	$1 \text{ in}^3 = \qquad \text{m}^3$	$1 \text{ m}^3 = \qquad \text{in}^3$

Notes: (a) the second column is the inverse of the first:

(b) The conversion between m^3 and cubic inches begins with the definition of an inch:

$$1 \text{ in} = 0.0254 \text{ m} \quad \text{(exactly)}$$

so

$$1 \text{ in}^3 = (0.0254)^3 \text{ m}^3$$

Exercise 5.13

Use this equation to complete the conversion table above

(Answer: $1 \text{ in}^3 = 1.639 \times 10^{-5} \text{ m}^3$; $1 \text{ m}^3 = 6.102 \times 10^4 \text{ in}^3$)

Example 5.9

A cube of marble has length 2 inches, calculate the volume in m^3

First, we can calculate the volume in cubic inches,

$$V = 2\text{ in} \times 2\text{ in} \times 2\text{ in} = 8\text{ in}^3$$

Now substitute for in^3 from the table

$$V = 8 \times (1.639 \times 10^{-5}\text{ m}^3) = 1.311 \times 10^{-4}\text{ m}^3.$$

Exercise 5.14

(a) A reactor has a volume of 6.7 mm^3, express the volume in m^3

$V = 6.7\text{ mm}^3 =$

(b) Express the volume $V = 100\text{ cm}^3$ in m^3

(c) Convert $V = 27\text{ in}^3$ into m^3

(d) A flask has a volume $3 \times 10^{-5}\text{ m}^3$, what is the volume in (i) cm^3 and (ii) dm^3

(i)

(ii)

(Answers, (a) $V = 6.7 \times 10^{-9}\text{ m}^3$; (b) $V = 1 \times 10^{-4}\text{ m}^3$; (c) $V = 4.42 \times 10^{-4}\text{ m}^3$; (d) (i) $V = 30\text{ cm}^3$, (ii) $V = 0.03\text{ dm}^3$)

Part (d) in the above exercise requires the use of the 'from m^3 column, replacing m^3 by the appropriate factor and required units.

(ii) converting between concentration units

When changing the units of concentration, we may have to deal with conversions of either the amount of substance or the volume (or both).

Experimentally, it is often convenient to report concentrations in mol dm^{-3}'(also abbreviated to M).

To convert from mol dm^{-3} to mol m^{-3}, we can use our knowledge from the previous section.
A difference here, though, is that the volume term occurs in the denominator ($c = n/V$).

As $1\text{ dm}^3 = 10^{-3}\text{ m}^3$, taking the inverse of each side gives

$$1\text{ dm}^{-3} = 10^{+3}\text{ m}^{-3}$$

i.e. there is a factor of 1000 in the conversion.

As a cubic metre is 1000 times larger than a cubic decimetre (litre), we get 1000 times the amount of substance in 1 m^3 than at the same concentration in 1 dm^3.

To convert from a concentration c quoted or measured in mol dm^{-3} to mol m^{-3} we simply replace dm^{-3} by 10^3 m^{-3} from above:

Example 5.10

Convert the concentration c = 0.1 mol dm^{-3} into mol m^{-3}:

$$c = 0.1 \text{ mol dm}^{-3} = 0.1 \text{ mol} \times (10^3 \text{ m}^{-3}) = 100 \text{ mol m}^{-3}$$

Conversion factors

to mol m^{-3}	from mol m^{-3}
1 M = 10^3 mol m^{-3}	1 mol m^{-3} = 10^{-3} M
1 mol dm^{-3} = 10^3 mol m^{-3}	1 mol m^{-3} = 10^{-3} mol dm^{-3}
1 mol L^{-1} = 10^3 mol m^{-3}	1 mol m^{-3} = 10^{-3} mol L^{-1}
1 mol cm^{-3} = 10^6 mol m^{-3}	1 mol m^{-3} = 10^{-6} mol cm^{-3}
1 mol mm^{-3} = 10^9 mol m^{-3}	1 mol m^{-3} = 10^{-9} mol mm^{-3}

Exercise 5.15

Convert the following concentrations to mol m^{-3}:

(a) c = 0.02 M

(b) $c = 3 \times 10^{-6}$ mol dm^{-3}

(c) c = 0.0005 mol cm^{-3}

(d) Convert c = 37.5 mol m^{-3} into M units

(Answers: (a) c = 20 mol m^{-3}; (b) $c = 3 \times 10^{-3}$ mol m^{-3}; (c) c = 500 mol m^{-3}; (d) c = 0.0375 M)

For gas-phase reactions, atmospheric and combustion chemists frequently choose to report concentrations in 'molecules per cubic centimetre', i.e. molecule cm^{-3}

This requires both a conversion of the volume and of the amount of substance.

To convert from molecule cm^{-3} to mol m^{-3}:

> We convert the volume term, replacing cm^{-3} by 10^6 m^{-3},
>
> then we convert the amount of substance by *dividing by* the Avogadro constant, $N_A = 6.022 \times 10^{23}$ molecule mol^{-1}

Note: it is helpful to write N_A with the 'unit' molecule as well as with mol^{-1} in this case.

We *divide* by N_A: this is consistent with the units and also logically consistent as we are converting from many molecules to few moles.

Example 5.11

A concentration of OH radicals is reported to be 3×10^{12} molecules cm^{-3}, express this concentration in mol m^{-3}

$$c = \frac{3\times10^{12}\,\text{molecule cm}^{-3}}{6.022\times10^{23}\,\text{molecule mol}^{-1}} = \frac{3\times10^{12}\,\text{molecule}\times(10^{6}\,\text{m}^{-3})}{6.022\times10^{23}\,\text{molecule mol}^{-1}} = 5\times10^{-6}\,\text{mol m}^{-3}$$

Overall the conversion factor between these units is

$$1 \text{ molecule cm}^{-3} = 1.66 \times 10^{-18} \text{ mol m}^{-3}$$

or

$$1 \text{ mol m}^{-3} = 6.022 \times 10^{17} \text{ molecule cm}^{-3}$$

Exercise 5.16

Convert the following concentrations between the units of mol m^{-3} and molecule cm^{-3}:

(a) $c = 5.2 \times 10^{13}$ molecule cm^{-3}

(b) $c = 0.05$ mol m^{-3}

(Answers: (a) $c = 8.64 \times 10^{-5}$ mol m^{-3}; (b) $c = 3 \times 10^{16}$ molecule cm^{-3})

(iii) converting between pressure units

The two 'empirical units' mmHg and Torr are defined as follows:

$$1 \text{ mmHg} = 133.322 \text{ Pa}$$
$$1 \text{ Torr} = (101\,325.0/760) \text{ Pa}$$

Numerically, these are virtually identical. 1 mmHg is related to the pressure exerted by a column of mercury 1 mm high under specified conditions of temperature, pressure and gravity, but is now best remembered in terms of the above numerical definition.

A related unit of mmH_2O, related to the pressure exerted by a column of water, in now defined as:

$$1 \text{ mmH}_2\text{O} = (133.322/13.5951) \text{ Pa}$$

Suggest an origin for the factor of 13.5951 in the expression above.

Example 5.12

The pressure in a reaction vessel is measured by a mercury manometer referenced to a vacuum. The heights of the two mercury columns are recorded as 100 and 532 mm respectively. Calculate the pressure and quote the result in SI units:

First, we can obtain the pressure in mmHg as the difference in height of the two mercury columns

$$p = 532 - 100 \text{ mmHg} = 432 \text{ mmHg}$$

next, we use the conversion factor from above:

$$p = 432 \times 133.322 \text{ Pa} = 57.595 \times 10^3 \text{ Pa}$$

or

$$p = 57.595 \text{ kPa}$$

Two other units for pressure commonly encountered are defined as:

standard atmosphere	$1 \text{ atm} = 101\,325 \text{ Pa}$
bar	$1 \text{ bar} = 100\,000 \text{ Pa} = 10^5 \text{ Pa}$
with, then,	
	$1 \text{ Pa} = (1/101\,325) \text{ atm} = 9.8692 \times 10^{-6} \text{ atm}$
	$1 \text{ Pa} = 10^{-5} \text{ bar}$

Example 5.13

Express the pressure from the last worked example in (a) atmospheres and (b) bar:

(a) $p = 57\,595 \text{ Pa} = (57\,595/101\,352) \text{ atm} = 0.56842 \text{ atm}$

(b) $p = 57\,595 \text{ Pa} = 57\,595 \times 10^{-5} \text{ bar} = 0.57595 \text{ bar}$

Exercise 5.17

Convert the following pressures to Pa:

(a) 100 mmHg

(b) 375 Torr

(c) 52.3 atm

(d) 250 bar

(e) convert the pressure $p = 1500$ Pa into: (a) mmHg; (b) Torr; (c) atmospheres and (d) bar

(Answers: p/Pa (a) 13332; (b) 49996; (c) 5.2993×10^6; (d) 25×10^6; (e) 11.25 mmHg, 11.25 Torr, 0.0148 atm, 0.015 bar)

Exercise 5.18

(a) Calculate the concentration and density of a sample of sulphur dioxide at 500 K and a pressure of 200 mmHg assuming it to behave as an ideal gas under these conditions:

$A_r(O) = 16, A_r(S) = 32$

(Hint: first convert the pressure to SI units, then use the formulae of Section 5.6)

(b) The molecule di-tertiary-butyl peroxide (DTBP: $(CH_3)_3COOC(CH_3)_3$) is compressed in a reactor of volume 0.25 dm^3 to a pressure of 5 atm at 500 K. Calculate the amount of substance present and hence the mass of the gas and its density.

(c) The DTBP compressed in the previous example then undergoes an irreversible reaction:

$$DTBP \rightarrow 2CH_3COCH_3 + C_2H_6$$

Calculate the new pressure in the system, the concentrations of acetone and ethane and the density of the product mixture.

(d) Bulb A of volume 50 cm^3 contains a gas under conditions of 27°C and 300 mmHg, and is separated from bulb B, of volume 5×10^{-2} L, containing gas at 177°C and 8×10^4 Pa by a closed tap. When the tap is opened and the gases allowed to mix the final pressure and temperature are 465 mmHg and 129°C. Are the gases behaving ideally on mixing ? [Take care in questions like this in which several different types of unit are involved.]

(Answers: (a) p = 26664 Pa, c = 6.41 mol m^{-3}, ρ = 0.411 kg m^{-3}; (b) n = 0.0305 mol, m = 4.45×10^{-3} kg, ρ = 17.8 kg m^{-3}; (c) p = 15 atm, $[CH_3COCH_3]$ = 0.244M, $[C_2H_6]$ = 0.122M, ρ = 17.8 kg m^{-3}; (d) No. If gas 1 were allowed to occupy the 100 cm^3 alone at the final temperature its final pressure would be 201 mmHg; similarly for gas 2 alone the final pressure would be 268 mmHg. Therefore, the calculated final pressure of the two gases together if behaving ideally would be (201 + 268) mmHg = 469 mmHg. This is greater than the actual measured final pressure and implies non-ideal behaviour.)

Important Equations used in this Section

The following should be familiar after completing this section and are gathered here for reference:

ideal gas equation $pV = nRT$

molar volume $V_m = \frac{V}{n}$

molar volume, ideal gas $V_m = \frac{RT}{p}$

concentration $c = \frac{n}{V}$

concentration, ideal gas $c = \frac{p}{RT}$

density $\rho = \frac{m}{V} = \frac{Mn}{V} = Mc$

mass of sample $m = Mn = McV = M\frac{pV}{RT}$

volume conversions

$1\ \mathrm{L} = 1 \times 10^{-3}\ \mathrm{m}^3$ $1\ \mathrm{m}^3 = 10^3\ \mathrm{L}$

$1\ \mathrm{dm}^3 = 1 \times 10^{-3}\ \mathrm{m}^3$ $1\ \mathrm{m}^3 = 10^3\ \mathrm{dm}^3$

$1\ \mathrm{cm}^3 = 1 \times 10^{-6}\ \mathrm{m}^3$ $1\ \mathrm{m}^3 = 10^6\ \mathrm{cm}^3$

$1\ \mathrm{mm}^3 = 1 \times 10^{-9}\ \mathrm{m}^3$ $1\ \mathrm{m}^3 = 10^9\ \mathrm{mm}^3$

concentration conversions

$1\ \mathrm{M} = 10^3\ \mathrm{mol\ m^{-3}}$ $1\ \mathrm{mol\ m^{-3}} = 10^{-3}\ \mathrm{M}$

$1\ \mathrm{mol\ dm^{-3}} = 10^3\ \mathrm{mol\ m^{-3}}$ $1\ \mathrm{mol\ m^{-3}} = 10^{-3}\ \mathrm{mol\ dm^{-3}}$

$1\ \mathrm{mol\ L^{-1}} = 10^3\ \mathrm{mol\ m^{-3}}$ $1\ \mathrm{mol\ m^{-3}} = 10^{-3}\ \mathrm{mol\ L^{-1}}$

$1\ \mathrm{mol\ cm^{-3}} = 10^6\ \mathrm{mol\ m^{-3}}$ $1\ \mathrm{mol\ m^{-3}} = 10^{-6}\ \mathrm{mol\ cm^{-3}}$

$1\ \mathrm{mol\ mm^{-3}} = 10^9\ \mathrm{mol\ m^{-3}}$ $1\ \mathrm{mol\ m^{-3}} = 10^{-9}\ \mathrm{mol\ mm^{-3}}$

$1\ \mathrm{molecule\ cm^{-3}} = 1.66 \times 10^{-18}\ \mathrm{mol\ m^{-3}}$ $1\ \mathrm{mol\ m^{-3}} = 6.022 \times 10^{17}\ \mathrm{molecule\ cm^{-3}}$

pressure conversions

1 mmHg = 133.322 Pa

1 Torr = (101 325.0/760) Pa

1 atm = 101 325 Pa

1 bar = 100 000 Pa = 10^5 Pa

1 Pa = (1/101325) atm = 9.8692×10^{-6} atm

1 Pa = 10^{-5} bar

SECTION 6

Graph Craft

Graph Craft

Graphs are used very widely in chemistry and other areas. They are excellent for summarising large amounts of information and presenting data economically. Experimental or computer-generated data can also be plotted in appropriate ways to test theories and allow other physical quantities to be determined.

6.1 Reading graphs

The graph below shows the *phase diagram* for a typical molecule. For different combinations of pressure p and temperature T, the molecule can exist as a gas, a liquid or a solid, with the p–T regions of these different *phases* indicated. Separating these regions are the *phase transition curves*, corresponding to *sublimation* (separating solid and gas), *fusion* or melting (separating solid and liquid) and *vaporisation* (liquid to gas).

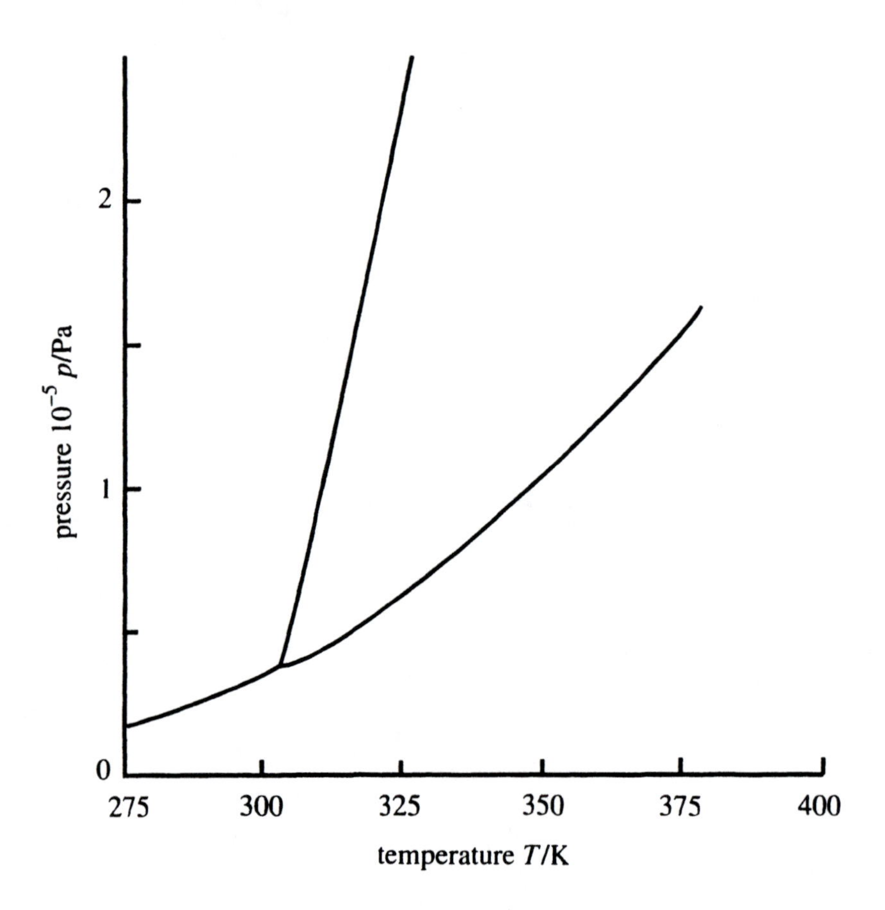

Exercise 6.1

Indicate which of the regions corresponds to the different phases and identify the three different phase transition curves on the diagram above:

(The solid phase can be expected at the lowest temperature and highest pressure: the gas phase can be expected at high temperature and low pressure.)

Answer:

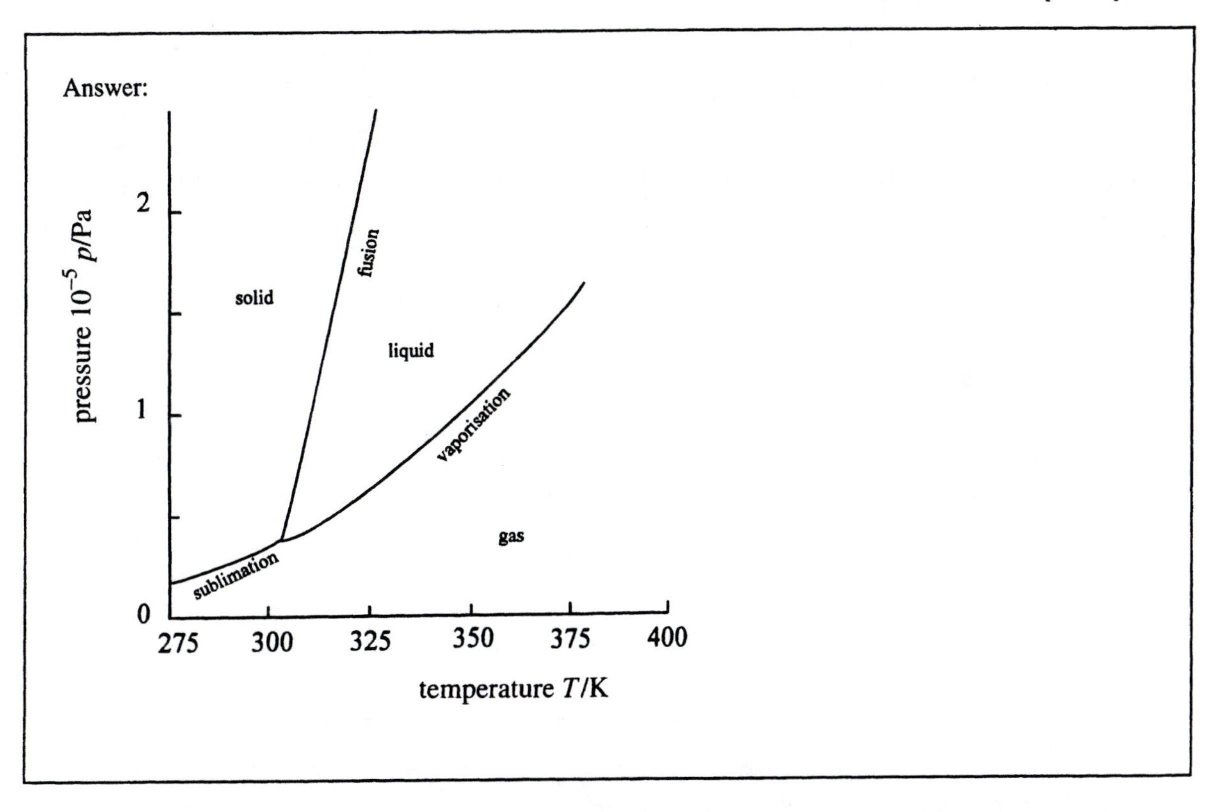

The axes have been labelled using the Guggenheim notation method (Section 3.5) with the appropriate SI units.

Exercise 6.2

(a) Determine which phase is found for the following conditions:

(i) $p = 1.5 \times 10^5$ Pa, $T = 300$ K

(ii) $p = 1.25 \times 10^5$ Pa, $T = 325$ K

(iii) $p = 0.5 \times 10^5$ Pa, $T = 350$ K

(iv) $p = 0.5$ atm, $T = 75°$C

(b) Estimate the standard melting point for this molecule, i.e. the melting point for the standard pressure $p^{\theta} = 1 \times 10^5$ Pa

(c) Estimate the standard boiling point for this molecule

(Answers: (a) (i) solid, (ii) liquid, (iii) gas, (iv) gas; (b) $T_m = 311$ K; (c) $T_b = 348$ K)

The 'normal' melting and boiling points correspond to a pressure of 1 atm (101 325 Pa) and are slightly higher than the standard values, although the difference will not be observable on the scale of the above figure.

There are two special points on the phase diagram. The *triple point* represents the p–T combination at which all three phases can co-exist at equilibrium. It is marked by the point at which the three phase transition curves meet.

The *critical point* represents the upper point at which the vaporisation curve terminates. For conditions beyond this point, the distinction between the liquid and gas phases disappears and the state is described as a *supercritical fluid*.

Exercise 6.3

Estimate the p–T locations of the triple and critical points.

(Answer: triple point $p_3 = 0.4 \times 10^5$ Pa, $T_3 = 303$ K; critical point $p_{cr} = 1.7 \times 10^5$ Pa, $T_{cr} = 379$ K)

The vaporisation curve can also correspond to the dependence of the vapour pressure p_{vap} established above the liquid in an open container on temperature (and the sublimation curve gives the variation of the vapour pressure above the solid phase).

Exercise 6.4

Estimate the vapour pressure at the following temperatures and indicate with which phase the vapour exists:

(a) 300 K

(b) 325 K

(c) 350 K

(Answers: (a) $p_{vap} = 0.36 \times 10^5$ Pa, solid; (b) $p_{vap} = 0.65 \times 10^5$ Pa, liquid; (c) $p_{vap} = 1.07 \times 10^5$ Pa, liquid)

A liquid boils when its vapour pressure becomes equal to the surrounding pressure.

Exercise 6.5

(a) At what pressure does the liquid have a boiling point of 350 K?

(b) What is the boiling point at a pressure of 0.5×10^5 Pa?

(Answer: (a) $p = 1.07 \times 10^5$ Pa; $T_b = 315$ K)

6.2 Plotting graphs

The second use of graphs is to test theoretical predictions and to use data to obtain the values of other physical quantities. As an example, we can use measured equilibrium constants to obtain values for the standard enthalpy and entropy changes for a reaction.

The dependence of the equilibrium constant K^θ for a chemical reaction on temperature is given by the equation

$$\ln\left(K^\theta\right) = -\frac{\Delta H^\theta}{R} \times \frac{1}{T} + \frac{\Delta S^\theta}{R}$$

where ΔH^θ and ΔS^θ are the standard enthalpy and entropy changes respectively.

In order to test the relationship, the reaction

$$N_2O_4(g) \rightarrow 2\ NO_2(g)$$

was studied at a range of temperatures. The following data were recorded:

T/K	K^θ
298	0.144
310	0.352
320	0.705
330	1.352
340	2.496
350	4.450
360	7.682
370	12.88

Exercise 6.6

Suggest a suitable straight line plot of these data to test this relationship

(Hint: compare the above equation with the standard form for a straight line $y = mx + c$ where m is the gradient and c the intercept)

(Answer: a plot of $\ln(K^\theta)$ versus $1/T$ would be appropriate)

Exercise 6.7

Explain how the values for ΔH^θ and ΔS^θ can also be obtained from this plot.

(Answer: gradient $m = -\Delta H^\theta / R$, intercept $c = \Delta S^\theta / R$)

Exercise 6.8

Calculate $\ln(K^{\theta})$ and $1/T$ for the first row in the table above

$\ln(K^{\theta})$ =

$1/T$ =

(Answer: $\ln(K^{\theta}) = -1.938$, $1/T = 3.356 \times 10^{-3}\ \text{K}^{-1}$)

The value of $1/T$ involves a factor of 10^{-3} and units of K^{-1} (i.e. inverse kelvin). This factor and unit will arise each time we calculate $1/T$ and so it is more convenient to use the Guggenheim notation method before tabulating and plotting these data.

We perform the following algebra with the power of 10 and the units:

First we write the equation in the form $\dfrac{1}{T} = 3.356 \times 10^{-3}\,\text{K}^{-1} = \dfrac{3.356}{10^3 \times \text{K}}$

then, we multiply the power of 10 and the unit up onto the left-hand side

$$\frac{10^3\,\text{K}}{T} = 3.356$$

This indicates that we should use $10^3\ \text{K}/T$ as our column heading and axis label.

Note that this equation can be re-arranged to give $10^3\ \text{K}/3.356 = T$, which gives $T = 298$ K when we calculate 1000/3.356.

Complete the following table from the results given previously

T/K	K^{θ}	10^3 K/T	$\ln(K^{\theta})$
298	0.144	3.356	−1.938
310	0.352		
320	0.705		
330	1.352		
340	2.496	2.941	0.915
350	4.450	2.857	1.493
360	7.682	2.778	2.039
370	12.88	2.703	2.556

Exercise 6.9

Comparing the theoretical relationship between $\ln(K^{\theta})$ and $1/T$, which quantity should be plotted along the x-axis and which along the y-axis?

x-axis:

y-axis:

(Answer: $1/T$ as x and $\ln(K^{\theta})$ as y)

In fact, we will plot $10^3\text{K}/T$ as the x-axis.

When plotting a graph, especially if we subsequently will wish to measure a gradient, it is important to use the paper size available to its maximum extent.

Exercise 6.10

Suggest suitable scales for the x- and y-axes on the graph below, bearing in mind the ranges covered by 10^3 K/T and ln(K^θ) in the table above.

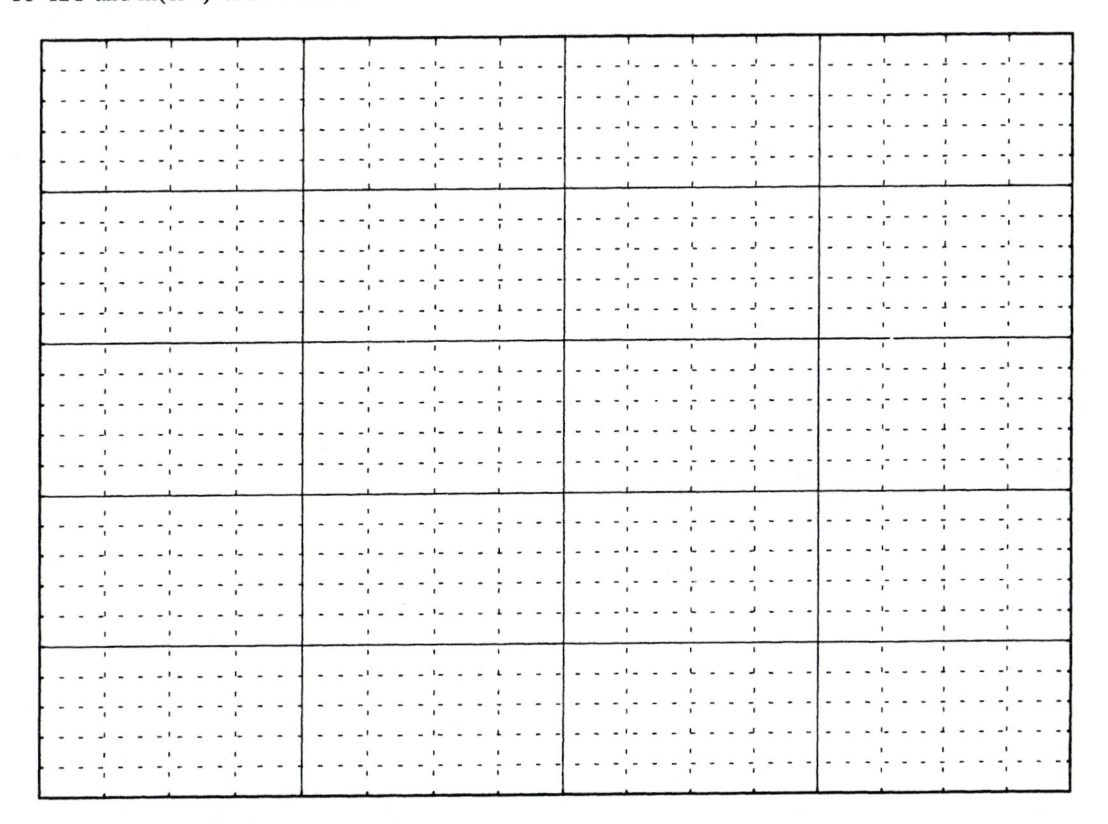

(Answer: sensible choices might be x-axis from 2.6 to 3.4 and y-axis from −2 to +3)

Exercise 6.11

Suggest the appropriate labels for the x- and y-axes

x-axis label:

y-axis label:

(Answers: x-axis, 10^3 K/T; y-axis, ln(K^θ))

Exercise 6.12

Plot the data points on the above graph and draw a straight line through the points

Estimate the gradient of the line

gradient =

Answer:

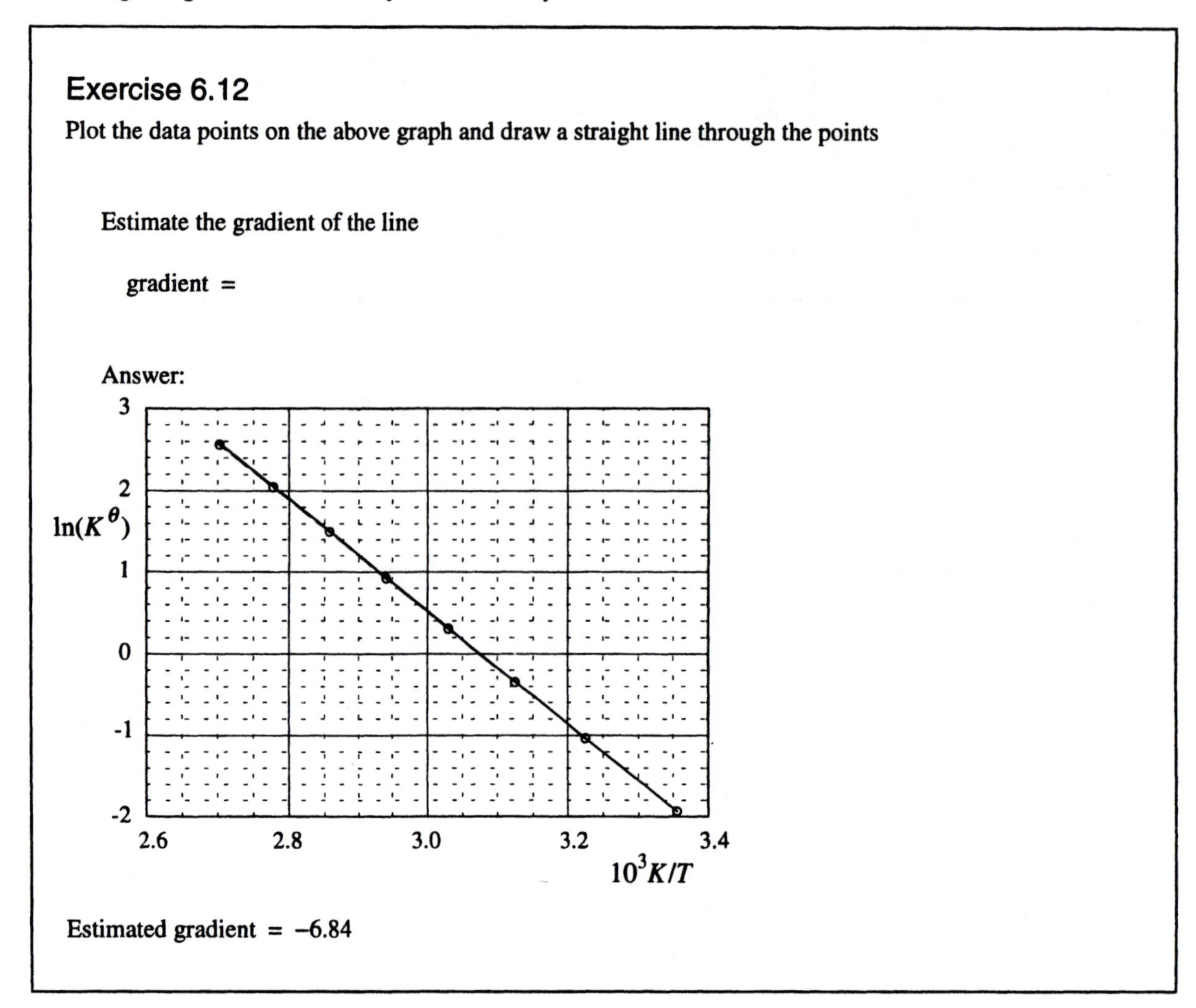

Estimated gradient = −6.84

When estimating the gradient, measure the changes in x and y over as wide a range of the graph as possible

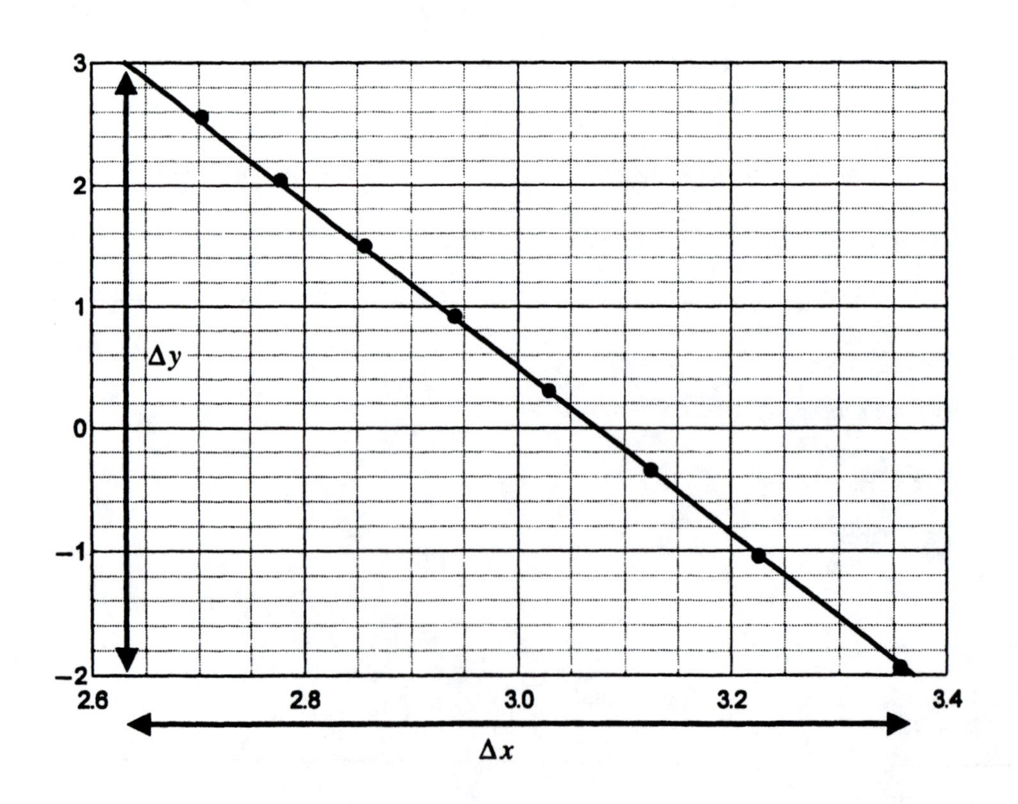

In this example, the whole range of the y-axis has been used, so $\Delta y = (-2) - 3 = -5$ units. Δx is then measured to be 106 mm long: noting that the total x-axis, corresponding to $3.4 - 2.6 = 0.8$ units, has a length of 116 mm, we calculate $\Delta x = 0.8 \times (106/116) = 0.73$. If we judge that we can measure these lengths to $\pm$ 0.5 mm, the uncertainty in each is approximately 0.5%, giving an uncertainty of 0.7% in Δx and, hence, in the calculated gradient.

(The uncertainty in Δx is obtained by combining the uncertainties in the two lengths according to the rule for propagating errors in a product or a quotient:

if $$x = a \times b \text{ or } x = \frac{a}{b}$$

and if the percentage uncertainties in a and b are $\%a$ and $\%b$ respectively, then the percentage uncertainty $\%x$ in x is calculated from

$$\%x = \sqrt{(\%a)^2 + (\%b)^2}$$

Using 0.5% for the uncertainty in both lengths in the previous example, we obtain the quoted uncertainty.)

Exercise 6.13

State the estimated gradient of the above line along with its absolute uncertainty

$m = -6.84 \pm$

(Answer: 0.7% of 6.84 is 0.05, so $m = -6.84 \pm 0.05$)

In general, the points will be scattered around some 'best fit' straight line, and a much larger uncertainty will arise in the gradient from this scatter than is caused by the precision with which the distances Δy and Δx can be measured if the above procedure of using the widest possible ranges on the graph is followed.

We have so far achieved one aim and are well placed for a second. First, the result that the points on the above graph do lie on a straight line acts as confirmation of the proposed relationship between $\ln(K^\theta)$ and $1/T$. Second, now that we have a value for the gradient, we can determine the standard enthalpy change ΔH^θ for the reaction.

One way to do this is to write down the appropriate form of the equation for a straight line, $y = mx + c$ for our particular graph.
We have plotted $\ln(K^\theta)$ along the y-axis and 10^3 K/T as the x-axis. From this graph, we have then determined $m = -6.84$.

Exercise 6.14

Substitute appropriately for y, x and m in the equation for a straight line for the graph (we still leave the intercept as c)

(Answer: $\ln\left(K^\theta\right) = -6.84 \times \dfrac{10^3\,\text{K}}{T} + c$: note that we must include the units and powers of 10 from the axes.)

We can now compare this form of the equation with that proposed at the beginning of this section:

From the graph $\ln(K^\theta) = \dfrac{-6.84\times10^3\,\mathrm{K}}{T} + c$

proposed $\ln(K^\theta) = \dfrac{-\Delta H^\theta}{RT} + \dfrac{\Delta S^\theta}{R}$

For these to be the same, we must have

$$\frac{-\Delta H^\theta}{R} = -6.84\times10^3\,\mathrm{K}$$

or, cancelling the – signs and re-arranging,

$$\Delta H^\theta = 6.84\times10^3\,\mathrm{K}\times R$$

Exercise 6.15

Use the value for R (including the correct units) to evaluate the standard enthalpy change for this reaction:

(Answer: $\Delta H^\theta = 6.84\times10^3\,\mathrm{K}\times 8.314\,\mathrm{J\,K^{-1}\,mol^{-1}} = 56.9\,\mathrm{kJ\,mol^{-1}}$)

Note how the factor 10^3 gives an answer in terms of kJ and the unit K from the x-axis cancels with the $\mathrm{K^{-1}}$ in R.

Returning to the comparison of the actual and proposed equations, the intercept c can now be related to the standard entropy change:

Exercise 6.16

How are c and ΔS^θ related?

(Answer: $c = \dfrac{\Delta S^\theta}{R}$)

The quantity c is the intercept of the straight line on the y-axis when $x = 0$. To determine the value of this intercept, we have several choices.

(i) We could redraw the graph on a scale that extends down to $x = 10^3\,\mathrm{K}/T = 0$ and then extrapolate the line through the points back to the y-axis. This approach, however, would suffer from the same problems of imprecision in estimating values from the graph described earlier. In the present case, also, it would involve a long extrapolation beyond the actual data points, and even small variations in the gradient will give rise to quite large changes in the extrapolated intercept.

(ii) We can choose a representative point along the line drawn on our existing graph, within the range of the actual data points and then use this and the measured gradient to determine c. This is a reasonable approach and it is best to choose a point somewhere near to the middle of the line through the data points. (The co-ordinates of the mid-point are likely to be least sensitive to small variations in the gradient.)

Exercise 6.17

Choose a point on the graph and estimate the x and y values:

x =

y =

Substitute these into the equation $y = mx + c$, along with the value determined previously for the gradient, $m = -6.84$, and then re-arrange this equation to obtain a value for the intercept c.

Use this result to evaluate ΔS^{θ}

(Answer: working from $x = 3.0$, $y = 0.50$ gives $0.50 = -6.84 \times 3.0 + c$, so $c = 0.50 + 6.84 \times 3.0 = 21.02$. Then, we use $c = \Delta S^{\theta} / R$ to obtain $\Delta S^{\theta} = c \times R = 21.02 \times 8.314 \text{ J K}^{-1} \text{ mol}^{-1} = 174.8 \text{ J K}^{-1} \text{ mol}^{-1}$)

The literature values for this reaction are $\Delta H^{\theta}_{298\text{K}} = 57.2 \text{ kJ mol}^{-1}$ and $\Delta S^{\theta}_{298\text{K}} = 176.9 \text{ J K}^{-1} \text{ mol}^{-1}$ respectively.

We have used the graphical plot of a set of experimental data to test a proposed relationship and determine the values of two physical quantities.

Exercise 6.18

The 'water gas shift' reaction is of great importance in the production of fuel by coal gasification as it relates to the interconversion of H_2 and CO. The reaction is

$$H_2(g) + CO_2(g) \rightleftharpoons H_2O(g) + CO(g)$$

The equilibrium constant K^{θ} has been measured at various temperatures as set out below.

Use this information to determine the standard enthalpy and entropy changes from an appropriate plot of the data

(The dependence of K^{θ} on T is the same as in the example above so the same approach can be used.)

T/K	K^{θ}
298	1.0×10^{-5}
500	7.76×10^{-3}
700	0.123
800	0.288
900	0.603
1000	0.955
1200	2.10

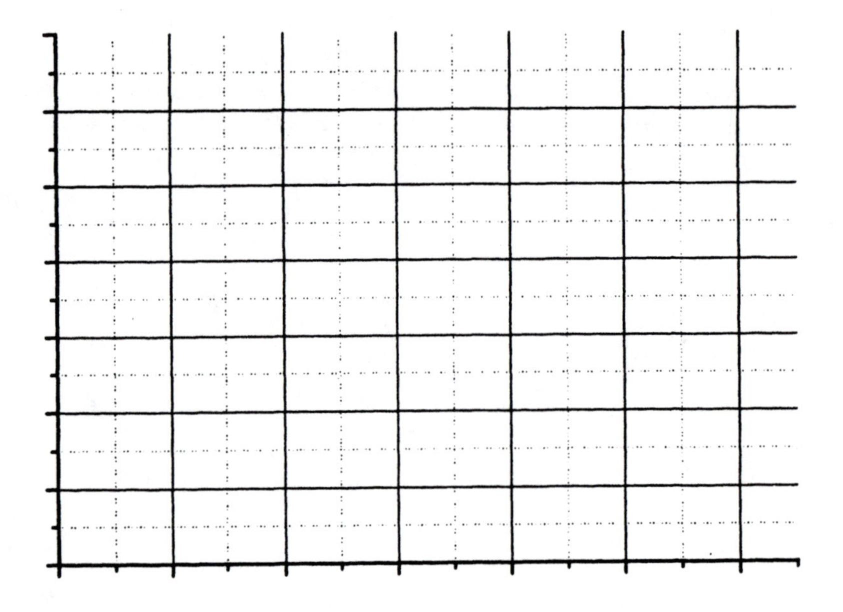

Use the gradient to calculate ΔH^{θ} and determine the intercept to obtain ΔS^{θ}

Answer:

Plotting $\ln(K^{\theta})$ versus 10^3 K/T,

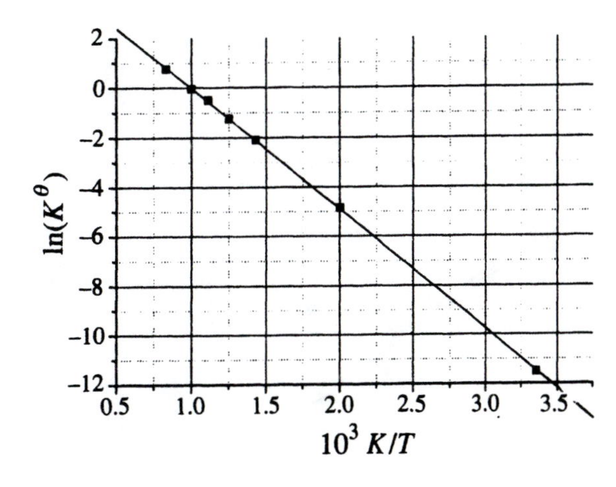

estimated gradient, $m = -4.87$ ΔH^{θ} = 40.5 kJ mol^{-1},
estimated intercept, $c = 4.85$ ΔS^{θ} = 40.3 J K^{-1} mol^{-1}

(literature values ΔH^{θ}_{298K} = 41.2 kJ mol^{-1}, ΔS^{θ}_{298K} = 42.4 J K^{-1} mol^{-1})

6.3 Least squares and linear regression

In the previous examples, the data points lie very much on a single straight line. In practice, experimental data are always gathered with some natural imprecision and frequently there is the additional problem of deciding on the 'best straight line' to fit through an appropriate plot. There are two approaches to this situation: a 'pragmatic' approach based on using the eye and a more rigorous, mathematical recipe that will determine explicitly the line of *least mean squares*.

Example 6.1

The following data were collected from a reaction kinetics experiment in which the concentration c of some reactant was determined at a series of times t.

time, t/s	concentration, c/M	$\ln(c/\text{M})$
85	0.136	−1.995
210	0.118	−2.137
315	0.0855	−2.459
395	0.0519	−2.958
500	0.0446	−3.110
620	0.0369	−3.300
700	0.0228	−3.781
805	0.0150	−4.200
907	0.0141	−4.262
1040	0.00985	−4.620

The reaction was run under so-called psuedo-first order conditions, so it is expected that a plot of $\ln(c)$ against t will be a straight line and that the gradient of this line will give the reaction rate constant. In fact, we expect

$$\ln(c) = \ln(c_0) - kt$$

where c_0 is the initial concentration.

The plot of $\ln(c/\mathrm{M})$ vs t/s is shown below.

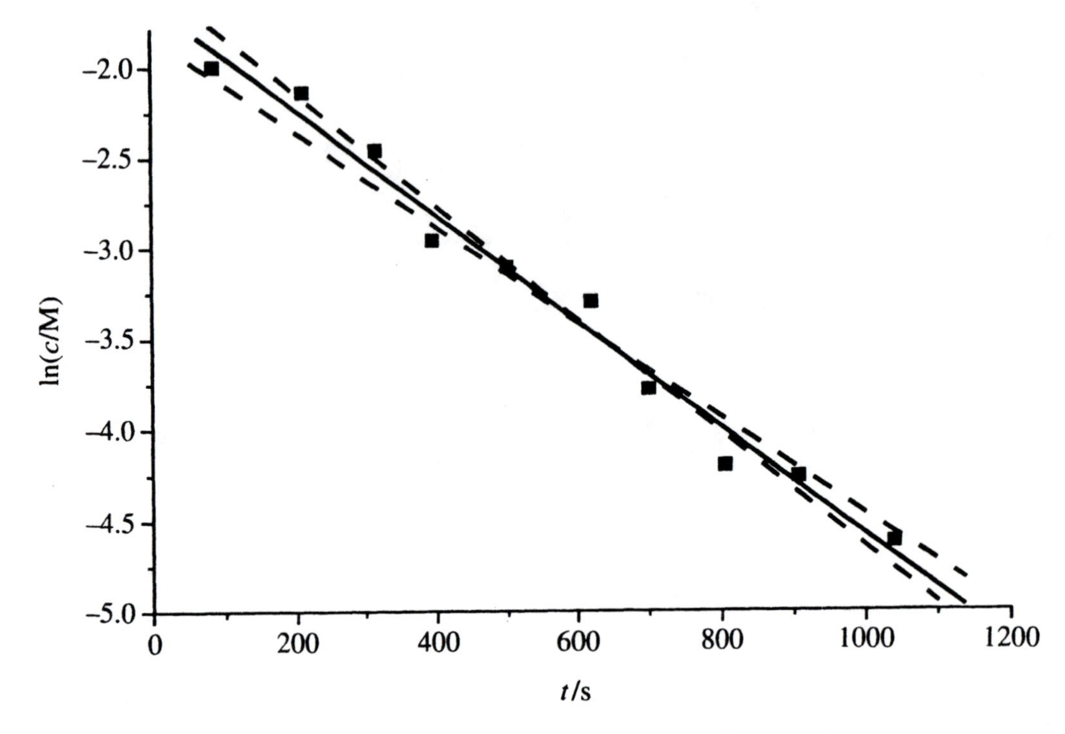

A 'best fit' straight line has been drawn on the basis of the positions of the experimental data points. The gradient m of this line is then estimated to be

$$m = -2.9 \times 10^{-3}$$

Using this value, and comparing the form $y = mx + c$ with the theoretical relationship above, we then obtain

$$k = 2.9 \times 10^{-3}\ \mathrm{s}^{-1}$$

(the units come because we have plotted t/s along the x-axis).

Also shown on the graph are two other straight lines (dashed). These represent a judgement of the steepest and least steep lines that still pass reasonably close to the majority of the data points. (They are obtained in practice by swivelling a ruler about some 'mid-point' of the data set.)

The slopes of these extreme fitted lines can be used to give an estimate of the uncertainty in the gradient. Again, measuring from the graph, the two gradients are estimated to be -3.3×10^{-3} and -2.5×10^{-3} respectively.

There is thus a difference of $\pm\, 0.4 \times 10^{-3}$, i.e. of approximately 14%, between the gradients of the two extreme lines and that of the estimated best fit line. This difference will lead to the same percentage difference in the calculated value of k, so we can quote our experimentally determined rate constant as

$$k = 2.9 \times 10^{-3}\,(\pm\,14\%)\ \mathrm{s}^{-1} \quad \text{or} \quad k = 2.9\,(\pm\,0.4) \times 10^{-3}\ \mathrm{s}^{-1}$$

The mathematical method of linear regression provides an explicit definition of what is the 'best' straight line for a given set of points. The gradient m and intercept c of the straight line are chosen so that the 'sum of the squares' of the difference between the actual values of y and the values of y calculated as $mx + c$ for each value of x is made as small as possible. In other words, for every point, we find the difference Δy between the actual y value y_{expt} and the calculated value y_{calc}. These numbers are then squared (this is done for two reasons: first so that they all become positive: otherwise a point lying a long way below the line might fortuitously cancel with another lying a long way above the line, and second so that points lying further away make a bigger contribution to the sum, which prevents the line being chosen to fit some points well but almost totally ignoring others). The actual sum of these squares will depend on the values chosen for m and c, and the least squares analysis ensures that these two important parameters are chosen so that the sum of the Δys is as small as possible.

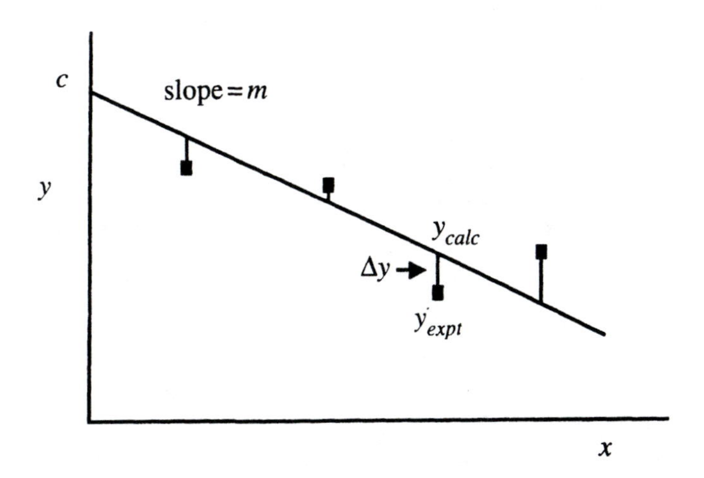

Schematic representation of least squares method: the difference $\Delta y = y_{expt} - y_{calc}$ is shown for one of the four points; this difference could be reduced by altering the gradient and intercept, but that would increase the differences for the other points. The best fit line makes the sum of all the $(\Delta y)^2$ values a minimum.

If each point is represented by its co-ordinates (x_i, y_i) with $i = 1$ to n for a set of n points, then the formulae for calculating the gradient m, intercept c and the variances α_m and α_c (estimates of the uncertainties Δm and Δc) are:

$$m = \frac{\sum_{i=1}^{n}(x_i - \bar{x})(y_i - \bar{y})}{\sum_{i=1}^{n}(x_i - \bar{x})^2}, \quad c = \bar{y} - m\bar{x}, \quad \alpha_m^2 = \frac{n\sigma^2}{n\sum x_i^2 - \left(\sum x_i\right)^2}, \quad \alpha_c^2 = \frac{\sigma^2 \sum x_i^2}{n\sum x_i^2 - \left(\sum x_i\right)^2}$$

with

$$\sigma^2 = \frac{\sum (y_i - mx_i - c)^2}{n - 2}$$

where $\bar{x}$ and $\bar{y}$ are the average values of x and y.

In practice, in modern chemistry laboratories, the linear regression fit is performed using computer-based graphics packages.

Example 6.2

For the data in the kinetics experiment discussed above, a linear regression fit produces the following graph

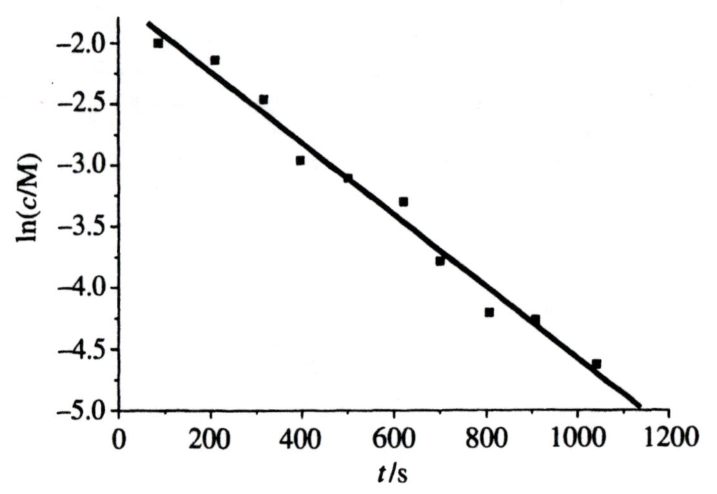

with the calculated gradient and intercept as follows:

$$m = -2.92\ (\pm 0.14) \times 10^{-3}$$
$$c = -1.65\ (\pm 0.09)$$

Noting that the intercept gives the value of $\ln(c_0/\mathrm{M})$, where c_0 is the initial concentration, these give

$$k = 2.92\ (\pm 0.14) \times 10^{-3}\ \mathrm{s}^{-1}$$
$$c_0 = 0.19\ (\pm 0.02)\ \mathrm{M}$$

(The uncertainty in c_0 can be determined by noting that we have to evaluate $e^{-1.65\pm0.09} = e^{-1.65} \times e^{\pm0.09}$. The second term is 1.094, indicating that there is an uncertainty of 9.4% in c_0.)

Exercise 6.19

(These are best performed using a computer-based graphics package if available.)

(a) The vapour pressure of water varies with temperature in the following manner.

T/K	273	283	293	303	313	323	333	343	353
p_{vap}/Pa	610	1230	2340	4240	7380	12 300	19 900	31 200	47 300

The dependence of p_{vap} on T, according to the Clausius–Clapeyron equation is

$$\ln\left(p_{vap}\right) = -\frac{\Delta H}{RT} + C$$

where C is some constant.

Suggest how the above data may be plotted to obtain the enthaply of vaporisation ΔH.

From this plot, determine ΔH and the appropriate uncertainty.

(b) The emission lines in the H atom spectrum that form the *Balmer Series* occur at the following wavenumbers $\overline{\nu}$

n	3	4	5	6	7
$10^{-3}\ \overline{\nu}\,/\,\mathrm{cm}^{-1}$	15.25	20.59	23.06	24.37	25.22

The corresponding quantum number n is indicated in the above table, and the relationship between n and $\overline{\nu}$ predicted by theory is

$$\overline{\nu} = \Re\left(\frac{1}{4} - \frac{1}{n^2}\right)$$

From the predicted gradient and intercept, explain how the Rydberg constant $\Re$ can be obtained from a plot of $\overline{\nu}$ against $1/n^2$.

Determine $\Re$ and the associated uncertainty.

Answers

(a) plot $\ln(p_{\mathrm{vap}}/\mathrm{Pa})$ vs $10^3\ \mathrm{K}/T$

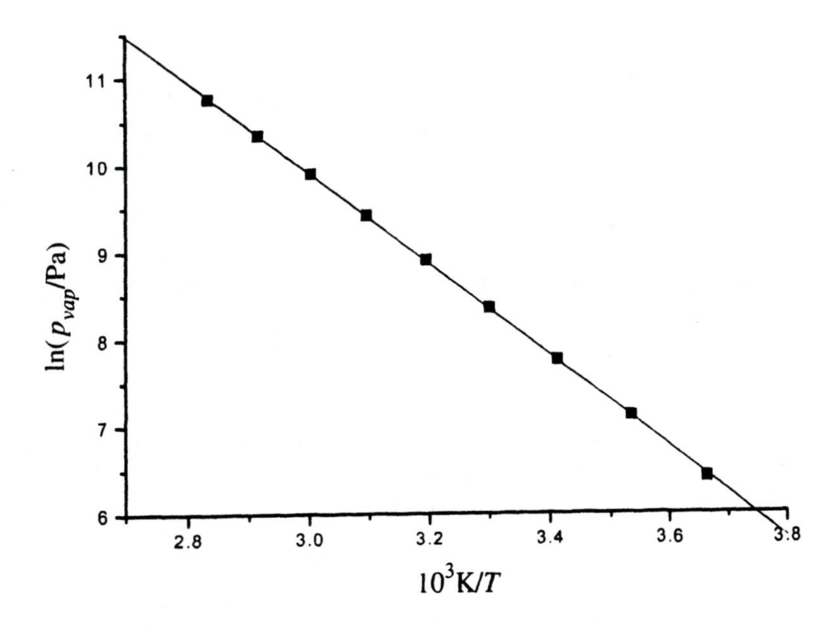

From linear regression analysis: gradient, m = −5.24 ± 0.02, giving $\Delta H = -m \times R \times 10^3\mathrm{K}$ = 43.6 (± 0.17) kJ mol^{-1}

(b) A plot of $\overline{\nu}$ against $1/n^2$ will have gradient $m = -\mathfrak{R}$ and an intercept $c = \mathfrak{R}/4$

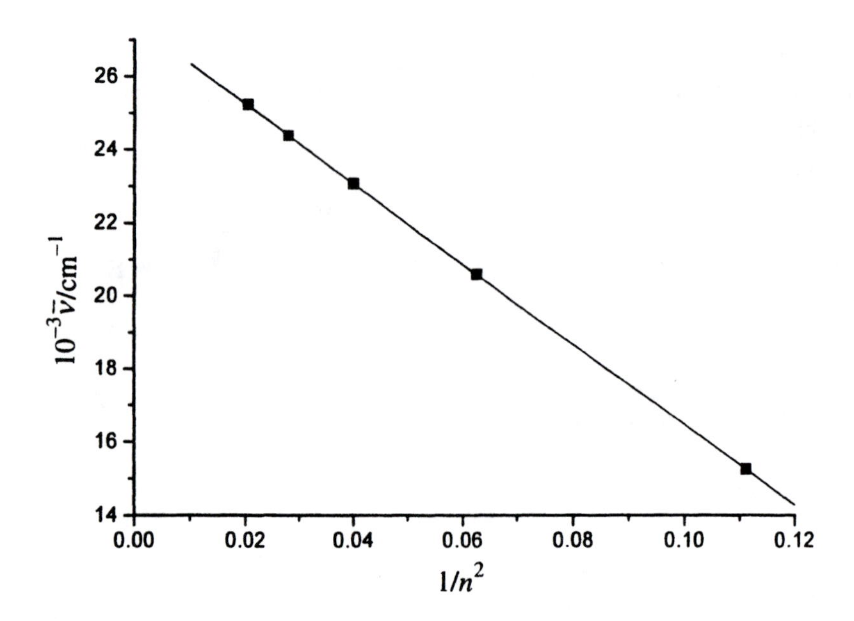

gradient $m = -109.73 \pm 0.24$, intercept $c = 27.443 \pm 0.015$, giving
from gradient: $\mathfrak{R} = 109730\ (\pm 240)\ \text{cm}^{-1}$
from intercept: $\mathfrak{R} = 109772\ (\pm 60)\ \text{cm}^{-1}$
Both give very precise results, with the intercept having the lower uncertainty.

6.4 Forcing plots through the origin

In some cases, theory will suggest that the relationship between two quantities corresponds to a special case of the equation for a straight line, namely to

$$y = mx$$

i.e. so that the intercept $c = 0$ and the line should then pass through the origin. If such a situation arises, then 'forcing' the linear regression analysis to produce a line passing through the origin can frequently produce more precise information.

The formula for the gradient of the least squares line passing through the origin is

$$m = \sum_{i=1}^{n} x_i y_i \Big/ \sum_{i=1}^{n} x_i^2$$

Example 6.3

Ferroin is a complex of Fe(II) ions with the ligand 1,10-phenanthroline and dissolves in water to give a deep red solution. The maximum absorbance of such solutions occurs at a wavelength $\lambda = 500$ nm. The variation of the absorbance A at this wavelength with concentration c was measured in a cell of path length $l = 1$ cm to establish whether the Beer–Lambert law is obeyed.

The Beer–Lambert law predicts

$$A = \varepsilon lc$$

where ε is the molar (decadic) extinction coefficient.

The following data were obtained.

$10^3\ c$/M	0.010	0.025	0.050	0.075	0.100
A	0.125	0.218	0.475	0.738	0.974

Plotting A versus $10^3\ c$/M, gives the following graph

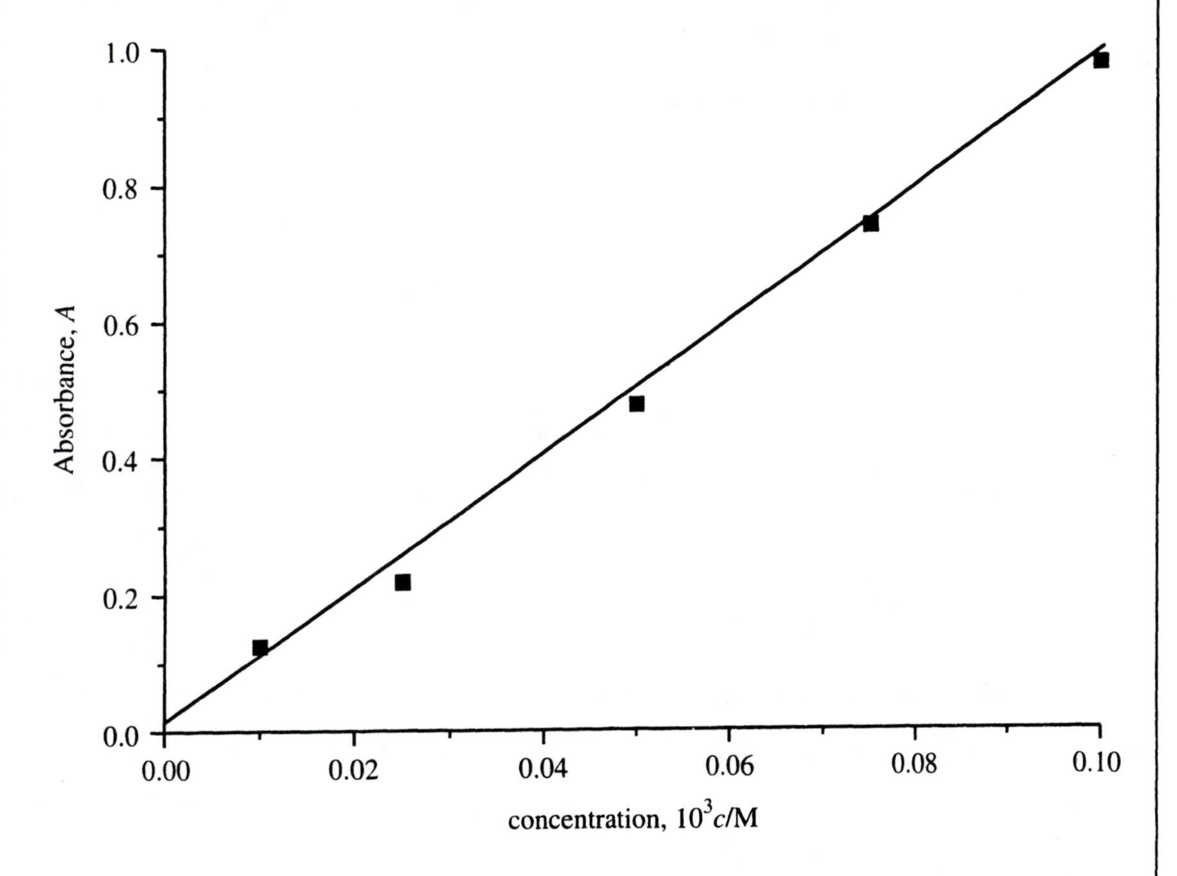

The linear regression analysis then gives: gradient $m = 9.68 \pm 0.32$, intercept $C = 0.002 \pm 0.019$.

For the Beer–Lambert law to hold, we expect a linear relationship and also, in this case, that the line should pass through the origin. The raw results do not meet the second requirement precisely and we need to be able to judge whether the non-zero value reported for C is significant or not.

The important question is whether the non-zero value for the intercept is small or large compared with the uncertainty in that quantity. In the present case, the uncertainty is $\pm$ 0.019, which is a factor of almost 10 (an 'order of magnitude') bigger than the apparent value of C. In this case then, we can conclude that the intercept is **not** significantly different from zero, and this part of the Beer–Lambert prediction is also satisfied.

From the above equation and the calculated gradient we can determine ε. Remembering the factor and units in the x-axis we obtain

$$\varepsilon l = 9.684\ (\pm 0.316) \times 10^3/\text{M}$$

so, with $l = 1$ cm, we obtain

$$\varepsilon = 9684\ (\pm 316)\ \text{M}^{-1}\ \text{cm}^{-1}$$

This rather unusual (and non-SI) set of units is frequently encountered in this particular area of physical chemistry as it is convenient to use with concentrations measured in molar units and path lengths of 1 cm.

Most computer graphics and linear regression packages have the option of forcing a fit through the origin. It is important that this option is **not** chosen immediately. Instead, the procedure above should be followed to determine whether the intercept for the 'non-forced' fit is or is not significantly different from zero. Only once this has been tested should the 'fit through the origin' option be selected. (In other words, we should give the data the chance to 'fail' to pass through the origin if that is genuinely what it does, irrespective of how tidy the theory that predicts $C = 0$.)

In the above example, we have established that the data do conform to a fit through the origin and so, the gradient can be re-computed on this basis.

For the above data, this then gives

$$m = 9.72 \pm 0.15$$

leading to

$$\varepsilon = 9720\ (\pm 150)\ \mathrm{M^{-1}\ cm^{-1}}$$

The fit through the origin results in a significantly lower value for the uncertainty.

(**Note**: simply adding the point (0,0) to a data set does not automatically cause a computer-based package to force the best fit straight line through the origin—the origin will simply be treated as another point, close to which the line will need to pass. The 'fit through zero' option will use a modified form of the regression formulae to ensure $C = 0$.)

6.5 Is it a good straight line?

Statistical packages are useful tools for the chemist and can return regression parameters for almost any set of data. This does not always mean, however, that the output from the analysis is meaningful. We need also to be able to judge whether the data really does lie on a straight line (or other fitting function).

Example 6.4

Shown opposite is a plot against carbon number n of the observed melting and boiling points for a set of straight-chain alkanes of general formula C_nH_{2n+2} arising from the following data:

n	T_m/K	T_b/K
3	85.46	231.1
4	134.8	272.65
5	143.43	309.22
6	177.8	341.9
7	182.54	371.57
8	216.35	398.81
9	219.63	423.94
10	243.49	447.27

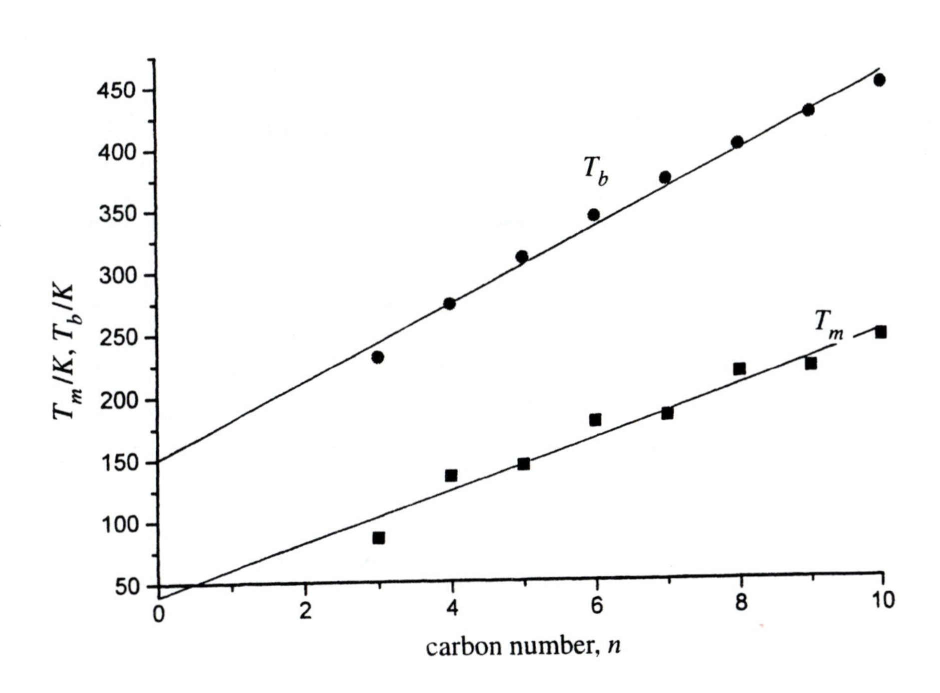

In each case, a linear regression analysis has been performed which results in the following formulae relating T_m and T_b to n:

$$T_m / K = 21(\pm 2)n + 38(\pm 12)$$

and

$$T_b / K = 31(\pm 1)n + 151(\pm 8)$$

The uncertainties in the gradient and intercept are quoted in each case, with the greatest uncertainty (approximately 32%) arising in the intercept for T_m.

The fact that the package has been able to obtain a least squares line and the associated fitting parameters does not, in itself, mean that the data genuinely do lie on a straight line, i.e. that the transition point temperatures genuinely are linearly related to carbon number.

We can, in part, use our own judgement and visual inspection of the graph to decide if there is a 'reasonable correlation' amongst each data set. A more quantitative test involves the *correlation coefficient*, r (often denoted as R in computer packages) which is calculated from the following formula:

$$r = \frac{\sum_{i=1}^{n}(x_i - \bar{x})(y_i - \bar{y})}{\sqrt{\sum_{i=1}^{n}(x_i - \bar{x})^2 \sum_{i=1}^{n}(y_i - \bar{y})^2}}$$

For data that lie on a perfect straight line, r will be equal to ± 1: if there is no correlation, $r = 0$.

Example 6.4 continued

For the melting and boiling point data, we obtain

r = 0.979 (melting point) and r = 0.995 (boiling point)

These numbers appear reasonably close to 1, but we have a relatively small sample. The way these parameters are used is in conjunction with a *significance test table*. We start by suggesting the hypothesis that the x and y data are not correlated, so that the true value for r (which would be obtained with an infinite data set) should be zero. The significance test table then gives values of r for different sample sizes, such that if the calculated value for r exceeds these values, this 'null hypothesis' can be rejected with a prescribed degree of confidence.

Example 6.4 continued

For a data set of eight points, the 5%, 1% and 0.1% confidence limits for r are:

5%	1%	0.1%
0.707	0.834	0.925

For each of the data sets presented, the actual value of r exceeds even the 0.1% confidence limit, so we can be more than 99.9% certain that there is some relationship between T_m or T_b and n

The value of r can be taken to confirm that T_m and T_b are related to n and that there is *some* linearity in the relationship. However, it is important that we do not come to rely solely on the numerical value of certain coefficients in this sort of situation. Looking at the actual data, there is certainly some suggestion of curvature in the boiling point data. To examine this further, we can use the fitted regression line to predict the melting and boiling points for some higher hydrocarbons.

Exercise 6.20

Predict the melting and boiling points for (a) dodecane ($C_{12}H_{26}$) and (b) eicosane ($C_{20}H_{42}$) on the basis of the linear regression fit. Try to estimate the uncertainty in these data.

(Answers: (a) T_m = 290 (±36) K, T_b = 523 (±20) K; (b) T_m = 458 (±52) K, T_b = 771 (±28) K)

We can compare these predictions with the experimental values

	T_m/K	T_b/K
dodecane	263	489.5
eicosane	310	616

So, even allowing for the generous uncertainties, there are significant failures in the predicted extrapolated transition point temperatures in these cases.

The correlation coefficient is certainly no substitute for simple inspection of the actual data.

Exercise 6.21

The rate constant k for the reaction

$$OH + H_2 \rightarrow H_2O + H$$

can be evaluated over a range of temperature, T.

T/K	$10^{-5}\ k/\mathrm{M}^{-1}\ \mathrm{s}^{-1}$
300	0.040
350	0.420
400	2.52
450	11.0
500	33.3
550	80.9
600	175
650	341
700	594

What is the value of the rate constant at 500 K?

(Answer: $k = 33.3 \times 10^5\ \mathrm{M}^{-1}\ \mathrm{s}^{-1}$)

These data can be plotted as a simple scatter graph and, although there is no particular reason to do so, a linear regression analysis can be performed:

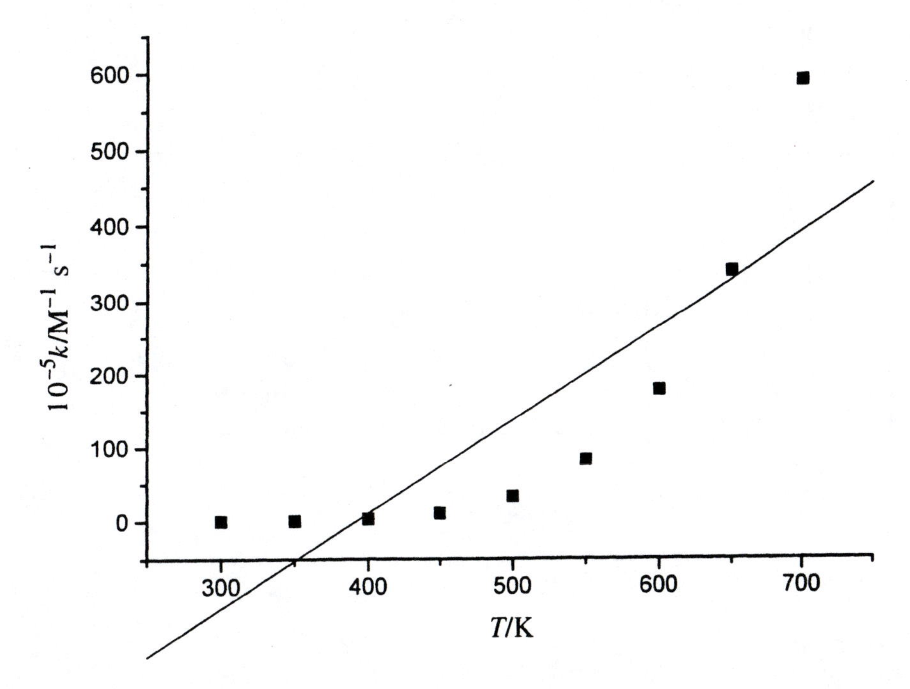

Inspection shows that the regression line is a very poor match to the actual data, but the correlation coefficient r has the value $r = 0.85$, which lies above the 99% confidence limit, indicating that there is some correlation between k and T.

The expected theoretical relationship between these data sets is the Arrhenius form

$$k = Ae^{-E/RT}$$

or

$$\ln k = \ln A - \frac{E}{RT}$$

where E is the *activation energy* and A is the *pre-exponential factor*.

Exercise 6.21 continued

Suggest a suitable plot of the data that should provide a linear relationship. How can A and E be determined in terms of the intercept and gradient?

(Answer: comparing with the general equation for a straight line $y = mx + c$, a plot of $\ln k$ as y against $1/T$ as x should be linear with an intercept $c = \ln A$ and a gradient $m = -E/R$)

Plot the data on the graph below

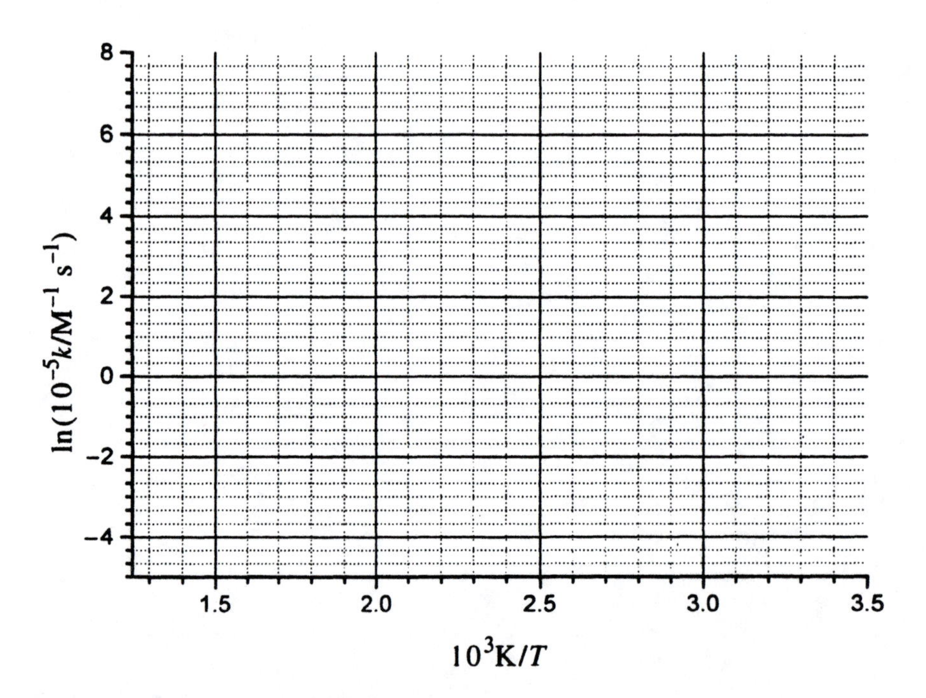

A regression analysis gives the following results:

gradient	$m = -5.05 \pm 0.01$
intercept	$c = 13.60 \pm 0.03$
correlation	$r = -0.99997$

Calculate the values of A and E.

(Answers: comparing the forms

$$\ln k = \ln A - \frac{E}{RT}$$

and

$$\ln\left(10^{-5}k / \mathrm{M^{-1}s^{-1}}\right) = 13.60(\pm 0.03) - \frac{5.05(\pm 0.01)\times 10^{3}\,\mathrm{K}}{T}$$

we have $\ln\left(10^{-5}A / \mathrm{M^{-1}s^{-1}}\right) = 13.60(\pm 0.03)$ so $10^{-5}A / \mathrm{M^{-1}s^{-1}} = e^{c} = e^{13.60\pm 0.03} = 8.1(\pm 0.3)\times 10^{5}$, i.e. $A = 8.1\ (\pm\ 0.3) \times 10^{10}\ \mathrm{M^{-1}\ s^{-1}}$ (notice that the units for A come from the units used for k); from the gradient, we have $-5.05\ (\pm\ 0.01) \times 10^{3}\mathrm{K} = -E/R$, giving $E = 41.99\ (\pm\ 0.08)\ \mathrm{kJ\ mol^{-1}}$.)

In this case, there is a very high correlation and the data clearly have a linear relationship in this form.

Exercise 6.22

In a chemical reaction, the concentration a of a reactant A is followed as a function of time t to give the following data:

t/s	a/M
1	0.909
2	0.833
3	0.769
4	0.706
5	0.667
6	0.618
7	0.594
8	0.556
9	0.520
10	0.500

We wish to determine the *order* of the reaction with respect to the concentration of A, i.e. we wish to determine the value of n in the *reaction rate equation* $-\mathrm{d}a / \mathrm{d}t = ka^{n}$, where k is the rate constant.

The following table indicates which particular plot of these data should be linear for different possible values of n and how the value of k is related to the gradient m:

	n	plot		rate constant
		as y	as x	
zero order	0	a	t	$-k$
first order	1	$\ln(a)$	t	$-k$
second order	2	$1/a$	t	k

These three plots are shown below along with the regression and correlation data.

Select the 'best' straight line fit and hence determine (a) the reaction order, (b) the reaction rate constant and (c) the initial concentration a_0.

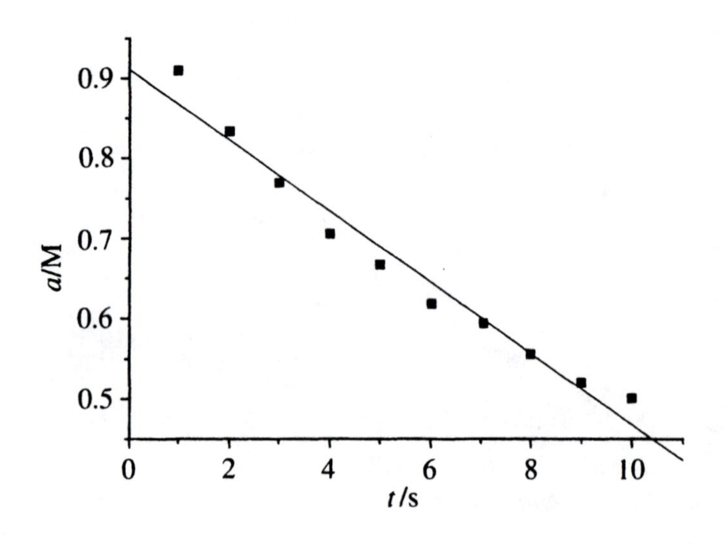
0.9
0.8
0.7
0.6
0.5
a/M
0
2
4
6
8
10
t/s

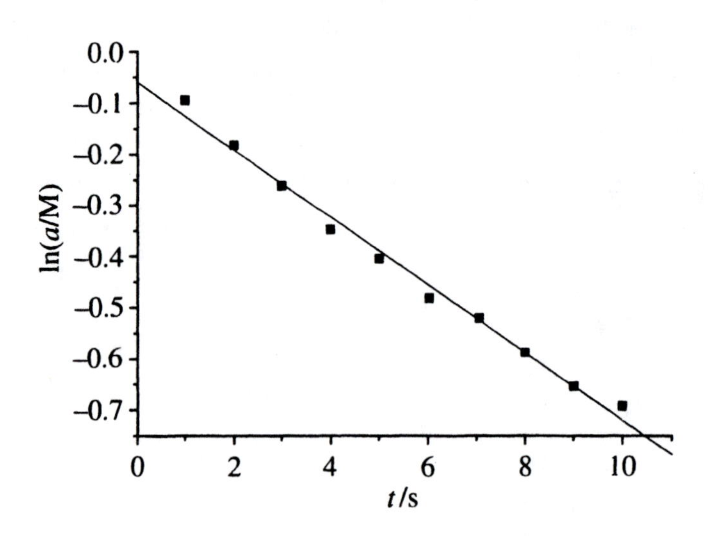
0.0
-0.1
-0.2
-0.3
-0.4
-0.5
-0.6
-0.7
ln(a/M)
0
2
4
6
8
10
t/s

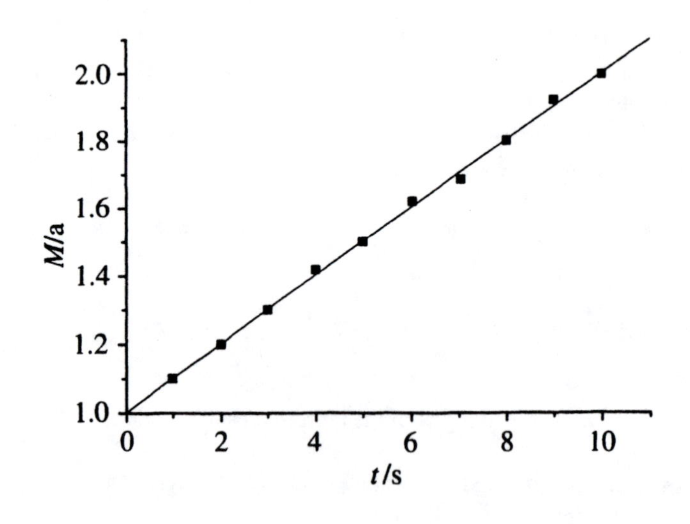
2.0
1.8
1.6
1.4
1.2
1.0
M/a
0
2
4
6
8
10
t/s

plot 1:	$m = -0.044\ (\pm 0.002)$,	$c = 0.91\ (\pm 0.02)$,	$r = -0.985$
plot 2:	$m = -0.066\ (\pm 0.002)$,	$c = -0.06\ (\pm 0.01)$,	$r = -0.996$
plot 3:	$m = 0.1004\ (\pm 0.001)$,	$c = 1.001\ (\pm 0.01)$,	$r = -0.999$

(Answers: (a) plot 3 has the highest correlation coefficient and is also visually the 'best fit', indicating that the reaction is second order, $n = 2$, with respect to a; (b) $k = 0.100\ \text{s}^{-1}$; (c) $a_0 = 1.00$ M)

Important Equations used in this Section

The following should be familiar after completing this section and are gathered here for reference:

equation of straight line $\quad y = mx + c \quad m$ = gradient, c = intercept

linear regression analysis

mean values

$$\bar{x} = \frac{1}{n}\sum_{i=1}^{n} x_i; \quad \bar{y} = \frac{1}{n}\sum_{i=1}^{n} y_i$$

least squares gradient

$$m = \frac{\sum_{i=1}^{n}(x_i - \bar{x})(y_i - \bar{y})}{\sum_{i=1}^{n}(x_i - \bar{x})^2},$$

least squares intercept

$$c = \bar{y} - m\bar{x};$$

variance

$$\sigma^2 = \frac{\sum (y_i - mx_i - c)^2}{n-2}$$

variance of mean

$$\alpha_m^2 = \frac{n\sigma^2}{n\sum x_i^2 - \left(\sum x_i\right)^2}$$

variance of intercept

$$\alpha_c^2 = \frac{\sigma^2 \sum x_i^2}{n\sum x_i^2 - \left(\sum x_i\right)^2}$$

linear regression through origin

$$m = \sum_{i=1}^{n} x_i y_i \Big/ \sum_{i=1}^{n} x_i^2$$

correlation coefficient

$$r = \frac{\sum_{i=1}^{n}(x_i - \bar{x})(y_i - \bar{y})}{\sqrt{\sum_{i=1}^{n}(x_i - \bar{x})^2 \sum_{i=1}^{n}(y_i - \bar{y})^2}}$$

The more specifically chemical equations appear elsewhere in this Workbook.

SECTION 7

Kinetic Theory of Gases

Kinetic Theory of Gases

Notation

The following notation will be used consistently through this section:

Symbol		SI unit
A_r	relative atomic mass	–
c	concentration	mol m^{-3}
C	heat capacity	J K^{-1}
C_m	molar heat capacity	J K^{-1} mol^{-1}
m	mass of one molecule or atom	kg
M	molar mass	kg mol^{-1}
M_r	relative molecular mass	–
$\mathfrak{m}$	total mass of system	kg
n	amount of substance	mol
N	number of molecules or atoms	–
N_A	Avogadro constant	mol^{-1}
$\mathfrak{n}$	number density	m^{-3}
p	pressure	Pa
T	thermodynamic temperature	K
V	volume	m^3
z	collision frequency	s^{-1}
Z	collision density	m^{-3} s^{-1}
ρ	density	kg m^{-3}

7.1 Number density

The number density η is defined as the number of molecules per unit volume:

$$\eta = \frac{N}{V}$$

η thus has units of m^{-3}.

For an ideal gas, the number density is related to the amount of substance n or the pressure by the following equations

$$\eta = \frac{N}{V} = \frac{nN_A}{V} = N_A \frac{p}{RT}$$

Example 7.1

Calculate the number density of an ideal gas at standard pressure and 298 K

$$\eta = N_A \frac{p}{RT} = 6.022\times10^{23}\,\text{mol}^{-1} \times \frac{10^5\ \text{Pa}}{8.314\,\text{J K}^{-1}\,\text{mol}^{-1} \times 298\text{K}} = 2.43\times10^{25}\,\text{m}^{-3}$$

Exercise 7.1

(a) Calculate the number density for Ar atoms at $p = 1$ Pa and $T = 500$ K

$$\eta = N_A \frac{p}{RT} =$$

(b) Calculate the number density for 0.01 mol of He in a 5 dm^3 flask:

$V = 5\ dm^3 =$ m^3

$\eta =$

(Answers: (a) $1.45 \times 10^{20}\ m^{-3}$; (b) $N = 0.01\ \text{mol} \times N_A$, $V = 5 \times 10^{-3}\ m^3$ so $\eta = 1.2 \times 10^{24}\ m^{-3}$)

Number densities are frequently quoted in the non-SI units of molecules cm^{-3}: the conversion between these was discussed in Section 3.7 (ii).

Exercise 7.2

(a) Convert the result from Exercise 7.1 (a) above into molecule cm^{-3} units

Compare this with the number densities for the following cases:

(b) the surface atmosphere of Mars which has a pressure of 0.006 atm and $T = 223$ K

(c) a molecular beam with pressure 1×10^{-6} Torr and $T = 20$ K

(d) ultra-high vacuum conditions with $p = 1 \times 10^{-11}$ Torr and $T = 300$ K

(Answers: (a) $\eta = 1.45 \times 10^{14}$ cm^{-3}; (b) $p = 608$ Pa, $\eta = 1.98 \times 10^{17}$ cm^{-3}; (c) $p = 1.33 \times 10^{-4}$ Pa, $\eta = 4.83 \times 10^{11}$ cm^{-3}; (d) $p = 1.33 \times 10^{-9}$ Pa, $\eta = 3.22 \times 10^{5}$ cm^{-3})

7.2 Molecular speeds

The *root mean square* molecular speed c_{rms} is given by

$$c_{rms} = \left(\frac{3k_B T}{m}\right)^{1/2}$$

Here m is the molecular mass (units = kg) and k_B is the Boltzmann constant (units = J K^{-1}).

This equation can be re-expressed in terms of the molar mass $M = N_A m$ (units = kg mol^{-1}) and the Gas constant $R = N_A k_B$ (units J K^{-1} mol^{-1}):

$$c_{\mathrm{rms}} = \left(\frac{3RT}{M}\right)^{1/2}$$

In neither case does the mass correspond to the *relative molecular mass M_r*.

Example 7.2

Calculate the root mean square speed of N_2 molecules at 500 K

$$c_{\mathrm{rms}} = \left(\frac{3k_BT}{m}\right)^{1/2} = \left(\frac{3\times 1.381\times 10^{-23}\,\mathrm{J\,K^{-1}}\times 500\,\mathrm{K}}{(28\times 10^{-3}/6.022\times 10^{23})\,\mathrm{kg}}\right)^{1/2} = 667\,\mathrm{m\,s^{-1}}$$

or

$$c_{\mathrm{rms}} = \left(\frac{3RT}{M}\right)^{1/2} = \left(\frac{3\times 8.314\,\mathrm{J\,K^{-1}\,mol^{-1}}\times 500\,\mathrm{K}}{28\times 10^{-3}\,\mathrm{kg\,mol^{-1}}}\right)^{1/2} = 667\,\mathrm{m\,s^{-1}}$$

In the first calculation, the molecular mass is obtained following the recipe from Section 2.3. In both cases, the mass must be expressed in kg.

The units of c correspond to m s^{-1} as J/kg = $m^2\,s^{-2}$ (see Section 4).

Exercise 7.3

Calculate the r.m.s. speed of Kr atoms at 1500 K

c_{rms} =

(Answer: using M_r(Kr) = 83.8, c_{rms} = 668 m s^{-1})

The *most probable speed* c^* and the mean speed $\bar{c}$ are given, respectively, by the formulae:

$$c^* = \left(\frac{2k_BT}{m}\right)^{1/2} = \left(\frac{2RT}{M}\right)^{1/2}$$

and

$$\bar{c} = \left(\frac{8k_BT}{\pi m}\right)^{1/2} = \left(\frac{8RT}{\pi M}\right)^{1/2}$$

Exercise 7.4

(a) Calculate the most probable and mean speeds for N_2 at 500 K

$c^* =$

$\bar{c} =$

(b) Compare these with the root mean square speed

(Answers: (a) $c^* = 545\ \mathrm{m\ s^{-1}}$, $\bar{c} = 615\ \mathrm{m\ s^{-1}}$; (b) $c^* < \bar{c} < c_{rms}$)

For a given gas, the three speeds always have the same ratio as the *RT/M* factor cancels, i.e.

$$c^*:\bar{c}:c_{rms} = (2)^{1/2}:\left(\frac{8}{\pi}\right)^{1/2}:(3)^{1/2} = 1.41:1.60:1.73$$

so c^* is approximately 12% smaller than $\bar{c}$, whilst c_{rms} is approximately 8% higher.

Molecular speeds increase as the square root of the temperature ($c \propto T^{1/2}$)
Molecular speeds decrease as the square root of the molecular mass ($c \propto M^{-1/2}$)

Exercise 7.5

(a) Calculate the ratio of the speeds of N_2 at 300 K and 3000 K

$$\frac{c_{rms}(300\,\mathrm{K})}{c_{rms}(3000\,\mathrm{K})} =$$

(b) Calculate the ratio of the speeds of N_2 and Ar at 500 K

$$\frac{c_{rms}(\mathrm{N_2})}{c_{rms}(\mathrm{Ar})} =$$

(Answers: (a) $(300/3000)^{1/2} = 0.316$; (b) $(28/40)^{-1/2} = 1.20$, so $c(\mathrm{Ar}) = 558\ \mathrm{m\ s^{-1}}$)

The ratio $c(300\,\mathrm{K})/c(3000\,\mathrm{K})$ is the same for any ideal gas.

The ratio $c(\mathrm{N_2})/c(\mathrm{Ar})$ is the same at any temperature.

7.3 Relative speed

The formulae above correspond to the speed of molecules relative to a fixed observer.

To calculate the frequency z with which molecules of two gases A and B collide, we need to calculate the *relative mean speed* $\bar{c}_{Rel}$ of the molecules with respect to each other.

This involves the *reduced mass* μ given by

$$\mu = \frac{m_A m_B}{m_A + m_B}$$

where m_A and m_B are the individual molecular masses (see Section 2.5).

Then, $$\bar{c}_{Rel} = \left(\frac{8k_BT}{\pi\mu}\right)^{1/2}$$

Exercise 7.6

Calculate the relative mean speed of an N_2 molecule and an Ar atom at 500 K

First calculate μ using $M_r(N_2) = 28$ and $M_r(Ar) = 40$.

$\mu =$

Next, substitute into the equation for $\bar{c}_{Rel}$

$\bar{c}_{Rel} =$

(Answers: $\mu = 2.735 \times 10^{-26}$ kg; $\bar{c}_{Rel} = 802$ m s^{-1})

For two molecules of the same type, e.g. two N_2 molecules, the reduced mass is simply one-half of the mass of an individual molecule $\mu = 1/2\ m$ and so $\bar{c}_{Rel} = 2^{1/2}\bar{c}$: in the following sections, the factor $2^{1/2}$ arises occasionally, originating from the use of this substitution.

Exercise 7.7

Calculate the relative mean speed of N_2 molecules at 500 K

$\mu =$

$\bar{c}_{\text{Rel}} =$

(Answer: $\mu = \frac{28 \times 28}{28 + 28} \times \frac{1 \times 10^{-3}\,\text{kg mol}^{-1}}{6.022 \times 10^{23}\,\text{mol}^{-1}} = 2.32 \times 10^{-26}\,\text{kg}$; $\bar{c}_{\text{Rel}} = 870\ \text{m s}^{-1}$)

7.4 Collisions between molecules: pure gases

The collision frequency gives the number of collisions per unit time in a sample of gas. It depends on the speed at which the molecules are moving and also on the *collision cross-section* $\sigma = \pi d^2$, where d is the *collision diameter*.

The situation is simplest for pure gases, rather than for mixtures, as collision cross-sections and diameters are tabluated for these. Some examples are given below:

gas	He	Ne	Ar	H_2	N_2	O_2	CO	CH_4	C_2H_6	C_6H_6
σ/nm^2	0.21	0.24	0.36	0.27	0.43	0.40	0.43	0.46	0.61	0.88

$1\ \text{nm}^2 = 1 \times 10^{-18}\ \text{m}^2$

In 1 s a molecule can be imagined to sweep out a cylinder of volume $V = \sigma\bar{c}_{\text{Rel}}$ as illustrated in the following figure:

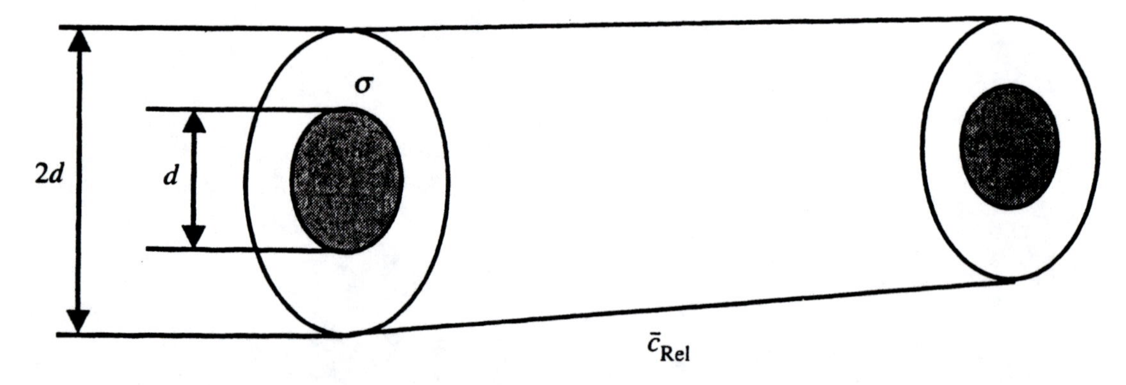

The molecule will collide with another molecule which has its centre lying inside this cylinder. The *collision frequency*, z, i.e. the number of collisions per second, is thus given by the product of the cylinder volume and the number density:

$$z = \sigma\bar{c}_{\text{Rel}}n$$

This result can be re-written in a number of other ways:

$$z = \sigma \bar{c}_{\text{Rel}} \mathcal{N} = \sigma \bar{c}_{\text{Rel}} \frac{N}{V} = \sigma \bar{c}_{\text{Rel}} \frac{p}{k_B T} = \sigma \bar{c}_{\text{Rel}} \frac{pN_A}{RT} = 2^{1/2} \sigma \bar{c} \frac{pN_A}{RT}$$

Exercise 7.8

Calculate the collision frequency for N_2 at 500 K at a pressure of 100 Pa, given $\sigma(\text{N}_2) = 0.43\ \text{nm}^2$

$z =$

(Answer: $z = 0.43 \times 10^{-18}\,\text{m}^2 \times 870\,\text{m s}^{-1} \times \dfrac{100\,\text{Pa}}{1.381 \times 10^{-23}\,\text{J K}^{-1} \times 500\text{K}} = 5.4 \times 10^6\,\text{s}^{-1}$)

At this low pressure a typical molecule makes over 5 million collisions per second.

The inverse of z gives the average time between collisions for a molecule.

Exercise 7.9

Calculate the average time between collisions for the N_2 molecules under the above conditions.

(Answer: time between collisions = $1/z = 0.185\ \mu\text{s}$)

7.5 Collision density: pure gases

The collision density Z gives the number of collisions per unit time per unit volume: it has units of $\text{s}^{-1}\ \text{m}^{-3}$.

For a single gas (rather than a mixture), the collision density Z_{AA} is given by

$$Z_{\text{AA}} = \tfrac{1}{2} z \mathcal{N} = \tfrac{1}{2} z \frac{N}{V}$$

where the extra factor of 1/2 arises to prevent each collision being counted twice.

Various other forms of this equation can be derived by substituting for z and re-writing in terms of either the concentration [A] or the pressure p

$$Z_{AA} = \tfrac{1}{2}\sigma\left(\frac{8k_BT}{\pi\mu}\right)^{1/2} N_A^2[A]^2 = \tfrac{1}{2}\sigma\left(\frac{8k_BT}{\pi\mu}\right)^{1/2} N_A^2\left(\frac{p}{RT}\right)^2 = \frac{1}{2^{1/2}}\sigma\left(\frac{8RT}{\pi M}\right)^{1/2} N_A^2\left(\frac{p}{RT}\right)^2$$

Note: here N_A is the Avogadro constant, as usual, not the number of A molecules (which is denoted by N). The final form uses the relationship between $\bar{c}_{Rel}$ and $\bar{c}$: M is the molar mass.

Example 7.3

Calculate the collision density for N_2 at 100 Pa

Using the penultimate form for Z_{AA}, we have

$$Z_{AA} = \tfrac{1}{2}\sigma\left(\frac{8k_BT}{\pi\mu}\right)^{1/2} N_A^2\left(\frac{p}{RT}\right)^2 =$$

$$\tfrac{1}{2}\times 0.43\times10^{-18}\,\mathrm{m}^2\times 870\,\mathrm{m\,s^{-1}}\times\left(6.022\times10^{23}\,\mathrm{mol^{-1}}\right)^2\times\left(\frac{100\,\mathrm{Pa}}{8.314\,\mathrm{J\,K^{-1}\,mol^{-1}}\times 500\,\mathrm{K}}\right)^2$$

$= 3.93 \times 10^{28}\ \mathrm{s^{-1}\ m^{-3}}$

At a given temperature, the collision density increases with the square of the pressure.
At a given pressure, the collision density decreases as the temperature increases.

Exercise 7.10

Explain the dependence of Z_{AA} on T:

$Z_{AA} \propto$

(Answer: $Z_{AA} \propto T^{-3/2}$: the collision density decreases because the concentration of the gas decreases more quickly than the mean speed increases as T changes)

At a fixed density, the collision number increases as the temperature increases.
The collision density decreases with the square root of the reduced mass of the molecules.

Exercise 7.11

(a) If $Z_{AA} = 2 \times 10^{30}$ s^{-1} m^{-3} for a sample of H_2 at 300 K, what would be the collision density for D_2 under the same conditions? (take $\sigma(D_2) \approx \sigma(H_2) = 0.27 \text{nm}^2$)

(b) What value would Z_{AA} take for these gases at 3000 K.

(c) The collision density is independent of size: true or false?

(d) If $Z_{AA} = 4 \times 10^{28}$ s^{-1} m^{-3} for a sample of N_2 what would be the collision density for C_2H_4 under the same conditions? ($\sigma(N_2) = 0.43 \text{nm}^2, \sigma(C_2H_4) = 0.57 \text{nm}^2$)

(Answers: (a) $Z_{AA}(D_2)/Z_{AA}(H_2) = [M(H_2)/M(D_2)]^{1/2} = 1/2^{1/2}$, so $Z_{AA}(D_2) = 1.4 \times 10^{30}$ s^{-1} m^{-3}; (b) $Z_{AA}(H_2) = 6.3 \times 10^{28}$ s^{-1} m^{-3}, $Z_{AA}(D_2) = 4.5 \times 10^{28}$ s^{-1} m^{-3}; (c) false: Z_{AA} depends on the size through the collision cross-section σ; (d) $Z_{AA}(C_2H_4) = 5.3 \times 10^{28}$ s^{-1} m^{-3})

7.6 Collisions for gas mixtures

For a mixture of two gases, the total collision frequency z is given by the same equation as before, with $\mathcal{n}$ being now the total number density. Of more interest in many situations, however, is the rate at which molecules of one type collide with molecules of another type, i.e. collisions of A with B.

The collision density Z_{AB}, giving the number of A–B collisions per second per unit volume, is given by the equation

$$Z_{AB} = \sigma \left(\frac{8k_BT}{\pi\mu} \right)^{1/2} N_A^2 [\text{A}][\text{B}]$$

where the collision cross-section $\sigma = \pi d_{AB}^2$ and d_{AB} is the mean molecular diameter $d_{AB} = \frac{1}{2}(d_A + d_B)$.

Exercise 7.12

The mean molecular diameter for a given gas can be calculated from the collision cross-section data σ given earlier. For example, for N_2 we have $\sigma = 0.43\ nm^2$. Using $\sigma = \pi d^2$, this leads to $d_{N_2} = 370\,pm$. Use the remaining σ data from the table in Section 7.4 to complete the following table:

gas	He	Ne	Ar	H_2	N_2	O_2	CO	CH_4	C_2H_6	C_6H_6
d/pm		276	339	293	370		370	383	441	529

(Answers: d(He)/pm = 259, $d(O_2)$/pm = 357)

Exercise 7.13

Calculate the average number of collisions per second between N_2 and O_2 molecules in air at 298 K and at a pressure of 1×10^5 Pa in a room of volume $V = 100\ m^3$. (Assume air has the composition $0.8N_2 + 0.2O_2$)

First calculate σ and μ

$\sigma =$

$\mu =$

Next, calculate $[N_2]$ and $[O_2]$ using $c = p/RT$

$[N_2] =$

$[O_2] =$

Then

$Z_{AB} =$

The collision frequency z is then given by

$z = Z_{AB} \times V =$

(Answers: $\sigma = 4.15 \times 10^{-19}\ m^2$, $\mu = 2.48 \times 10^{-26}$ kg; $[N_2] = 32.3\ mol\ m^{-3}$, $[O_2] = 8.07\ mol\ m^{-3}$; $Z_{AB} = 2.55 \times 10^{34}\ s^{-1}\ m^{-3}$; $z = 2.55 \times 10^{36}\ s^{-1}$)

7.7 Collisions with walls and surfaces

The number of collisions of the molecules in a gas with a surface per unit time and per unit area Z_w can be written in terms of the following equations

$$Z_w = \tfrac{1}{4}\bar{c}n = \left(\frac{k_B T}{2\pi m}\right)^{1/2} \frac{N}{V} = \left(\frac{RT}{2\pi M}\right)^{1/2} \frac{N}{V} = \frac{p}{(2\pi m k_B T)^{1/2}}$$

(The origin of the factor 1/4 in this equation involves the distinction between *speed* and *velocity*. The former just considers how fast a molecule is moving through space, but velocity also considers the specific *direction* in which the molecule is moving. Of the molecules lying sufficiently close to the wall to be able to hit it in one second, half will actually be travelling *away* from the wall and so will not make a hit. Of the remaining half, some will be travelling directly towards the wall, but the majority will be moving at some angle and so have further to travel before making a hit. Overall then, it is as though only 1/4 of the potential collisions per unit area actually happen.)

Exercise 7.14

Calculate the number of collisions per second z_w made by N_2 molecules at p = 100 Pa and T = 500 K with a surface of area 0.3 m^2.

Z_w =

$z_w = Z_w \times \text{area} =$

(Answer: $Z_w = 2.23 \times 10^{24}$ s^{-1} m^{-2}; $z_w = 6.68 \times 10^{23}$ s^{-1})

It is the collision of molecules with the walls of a container that give rise to the pressure exerted on the walls.

Exercise 7.15

(a) The reaction between CO and O_2 is catalysed by platinum metal surfaces. The maximum rate at which this catalysed reaction can occur is limited by the rate at which the reactants arrive at the surface. Calculate the maximum rate at which CO at a partial pressure of 1 × 10^3 Pa in air (corresponding approximately to 1% CO at atmospheric pressure) can be reacted per unit area of catalyst at 800 K. (Catalytic convertors in car exhaust systems use highly dispersed Pt crystallites to obtain high specific surface areas.)

(b) A typical value for the steady-state OH concentration in the stratosphere is 1.7×10^{-15} mol dm^{-3}. Calculate the rate at which OH radicals collide with ice particles of surface area $A = 1 \times 10^{-9}$ m^2 assuming a temperature of 240 K. (Reactions occurring on such particles are thought to be intimately involved in the production of acid rain.)

(Answers:

(a)

$$Z_w = 1\times10^3\,\text{Pa} \;/\; \left(2\pi\times\frac{28\times10^{-3}\ \text{kg mol}^{-1}}{6.022\times10^{23}\ \text{mol}^{-1}}\times1.381\times10^{-23}\,\text{J K}^{-1}\times800\,\text{K}\right)^{1/2} = 1.76\times10^{25}\ \text{m}^{-2}\,\text{s}^{-1},$$ this result can be converted into molar terms by dividing by N_A: $Z_w = 29.2$ mol m^{-2} s^{-1}.

(b) Noting that the concentration $c = 1.7 \times 10^{-15}$ mol dm^{-3} = 1.7×10^{-12} mol m^{-3} is equivalent to a number density $N/V = 1.02 \times 10^{12}$ m^{-3} and M(OH) = 17×10^{-3} kg mol^{-1}, we can use

$$Z_w = \left(\frac{8.314\,\text{J K}^{-1}\text{mol}^{-1}\times240\,\text{K}}{2\times\pi\times17\times10^{-3}\,\text{kg mol}^{-1}}\right)^{1/2}\times1.02\times10^{12}\,\text{m}^{-3} = 1.40\times10^{14}\ \text{m}^{-2}\,\text{s}^{-1}$$

giving 1.40×10^5 collisions s^{-1})

7.8 Mean free path

The mean distance travelled by a molecule between collisions is known as the *mean free path* λ. This can be obtained simply as the ratio of the mean speed to the collision frequency

$$\lambda = \frac{\bar{c}}{z}$$

If we are dealing with a pure gas A, of concentration [A], the equation can also be written as

$$\lambda = \frac{k_B T}{2^{1/2}\sigma p} = \frac{V}{2^{1/2}\sigma N} = \frac{1}{2^{1/2}\sigma N_A[\text{A}]}$$

Example 7.4

Ultra-high vacuum (u.h.v.) experiments are routinely performed with a total pressure of 1 nTorr, i.e. 1×10^{-9} Torr. What is the mean free path of N_2 molecues at 500 K under these conditions?

First, we convert the pressure to Pa using the method of Section 3.7:

$$p = 1 \times 10^{-9}\,\text{Torr} \times \frac{101325\,\text{Pa}}{760\,\text{Torr}} = 1.33 \times 10^{-7}\,\text{Pa}$$

Then, substituting into the above equation with $\sigma(N_2) = 0.43\ \text{nm}^2$, we have

$$\lambda = \frac{k_B T}{2^{1/2}\sigma p} = \frac{1.381 \times 10^{-23}\,\text{J K}^{-1} \times 500\,\text{K}}{2^{1/2} \times 0.43 \times 10^{-18}\,\text{m}^2 \times 1.33 \times 10^{-7}\,\text{Pa}} = 85.4\,\text{km}$$

This large mean free path suggests that under u.h.v. conditions, molecules rarely collide.

Exercise 7.16

(a) What pressure gives a mean free path of 0.1 m for N_2 at 500 K?

(b) What is the mean free path for N_2 at 500 K under 'ultra-high pressure' conditions with p = 1 Mbar?

$p =$ Pa

$\lambda =$

(c) In some u.h.v. experiments, a pressure as low as 1×10^{-11} Torr is achieved. Calculate the mean free path for C_2H_4 molecules at 298 K at this pressure.

(Answers: (a) $p = k_BT/2^{1/2}\sigma\lambda = 0.11$ Pa; (b) $\lambda = 1.14 \times 10^{-10}$ m; (c) 3.83×10^6 m)

Answer (a) suggests that for pressures less than 10^{-6} atm, the mean free path becomes comparable with the size of a 1 L flask.

Important Equations used in this Section

The following should be familiar after completing this section and are gathered here for reference:

number density
$$\eta = \frac{N}{V} = \frac{nN_A}{V} = N_A \frac{p}{RT}$$

molecular speeds
$$c_{rms} = \left(\frac{3k_BT}{m}\right)^{1/2} = \left(\frac{3RT}{M}\right)^{1/2} \quad \text{root mean square}$$
$$c^* = \left(\frac{2k_BT}{m}\right)^{1/2} = \left(\frac{2RT}{M}\right)^{1/2} \quad \text{most probable}$$
$$\bar{c} = \left(\frac{8k_BT}{\pi m}\right)^{1/2} = \left(\frac{8RT}{\pi M}\right)^{1/2} \quad \text{mean}$$

relative mean speed
$$\bar{c}_{Rel} = \left(\frac{8k_BT}{\pi\mu}\right)^{1/2} \qquad \mu = \frac{m_A m_B}{m_A + m_B}$$

for single gases $\bar{c}_{Rel} = 2^{1/2}\bar{c}$

collision frequency
$$z = \sigma\bar{c}_{Rel}\eta = \sigma\bar{c}_{Rel}\frac{N}{V} = \sigma\bar{c}_{Rel}\frac{p}{k_BT} = \sigma\bar{c}_{Rel}\frac{pN_A}{RT} = 2^{1/2}\sigma\bar{c}\frac{pN_A}{RT}$$

collision density
$$Z_{AA} = \tfrac{1}{2}z\eta = \tfrac{1}{2}z\frac{N}{V}$$
$$= \tfrac{1}{2}\sigma\left(\frac{8k_BT}{\pi\mu}\right)^{1/2} N_A^2[A]^2 = \tfrac{1}{2}\sigma\left(\frac{8k_BT}{\pi\mu}\right)^{1/2} N_A^2\left(\frac{p}{RT}\right)^2 = \frac{1}{2^{1/2}}\sigma\left(\frac{8RT}{\pi M}\right)^{1/2} N_A^2\left(\frac{p}{RT}\right)^2$$
$$Z_{AB} = \sigma\left(\frac{8k_BT}{\pi\mu}\right)^{1/2} N_A^2[A][B] \qquad d_{AB} = \tfrac{1}{2}(d_A + d_B)$$

collision with surface
$$Z_w = \tfrac{1}{4}\bar{c}\eta = \left(\frac{k_BT}{2\pi m}\right)^{1/2}\frac{N}{V} = \left(\frac{RT}{2\pi M}\right)^{1/2}\frac{N}{V} = \frac{p}{(2\pi m k_BT)^{1/2}}$$

mean free path
$$\lambda = \frac{\bar{c}}{z} = \frac{k_BT}{2^{1/2}\sigma p} = \frac{V}{2^{1/2}\sigma N} = \frac{1}{2^{1/2}\sigma N_A[A]}$$

SECTION 8

Chemical Thermodynamics

Chemical Thermodynamics

8.1 Pre-chemistry: work and heat

The work dw done by a gas expanding by a small volume dV against a pressure p is given by

$$dw = -p\,dV$$

If the gas is compressed, dV is negative and work is done on the gas.

In order to find the work done in a finite change, we need to integrate this equation.

There are two particularly common cases:

(i) An isothermal expansion against a fixed external pressure, p_{ex}, in which case

$$w = -p_{ex}\,\Delta V$$

where $\Delta V = V_{final} - V_{initial}$ is the (finite) volume change

(ii) An isothermal reversible expansion or compression in which the external pressure is continually adjusted to remain virtually equal to the internal pressure, then using $p = nRT/V$ and integrating gives

$$w = -nRT\,\ln\left(\frac{V_{final}}{V_{initial}}\right) = -nRT\,\ln\left(\frac{p_{initial}}{p_{final}}\right)$$

Example 8.1

Calculate the work done on 0.05 mol of an ideal gas when it is expanded from an initial volume of 0.025 m^3 to a final volume of 0.124 m^3 at 298 K (i) if the expansion is against a fixed external pressure of 1000 Pa and (ii) if the expansion is carried out reversibly

(i) $w = -p_{ex}\,\Delta V = -1000\text{ Pa} \times (0.124 - 0.025)\text{ m}^3 = -99\text{ J}$

(ii) $w = -nRT\ln(V_{final}\;/\;V_{initial})$

$= -0.05\text{ mol} \times 8.314\text{ J K}^{-1}\text{ mol}^{-1} \times 298\text{ K} \times \ln(0.124/0.025) = -198\text{ J}$

In each case, w is negative, indicating that the system has done work.

The internal energy, i.e. the capacity of the system to do further work, may also change during such expansions; however, the change ΔU in the internal energy involves both the work and the heat flow, which we address below.

Exercise 8.1

(a) Calculate the work done when 10^{-3} mol of Ar is expanded isothermally from 0.025 dm^3 to 0.1 dm^3 against a vacuum at 500 K

(b) Calculate the work performed if the above expansion is repeated against a pressure of 41.5 kPa

(c) Calculate the work done if this expansion is now performed adiabatically against a pressure of 41.5 kPa.

(d) Calculate the work performed when 10^{-3} mol of Ar is compressed isothermally and reversibly from a pressure of 50 kPa to a final pressure of 150 kPa at 500 K

(Answers: (a) $p_{ex} = 0$, so $w = 0$; (b) $\Delta V = 0.075 \times 10^{-3}$ m^3, $w = -$ 3.1 J; (c) $w = -$ 3.1 J ; (d) $w =$ 4.57 J)

During the above expansions (except in the adiabatic case), a quantity of heat q must be exchanged to maintain the constant temperature.

If this can be determined, then the total change in the internal energy, ΔU can be calculated from

$$\Delta U = q + w$$

There are, again, some common cases:

(i) for an isothermal process involving an ideal gas, the overall change ΔU is zero as the internal energy depends only on the temperature and not on the volume or pressure, so then

$$q = -w \qquad \textit{isothermal process}$$

(ii) for an adiabatic process, $q = 0$, i.e. there is no heat flow: this is the definition of adiabaticity

$$q = 0 \qquad \textit{adiabatic process}$$

(iii) for a process in which the temperature increases due to an exchange of energy as heat with the surroundings, then q is related to ΔT through the *heat capacity* of the system:

$$q = C_V \Delta T \text{ for a process at constant volume}$$
$$q = C_p \Delta T \text{ for a process at constant pressure}$$

Here C_V is the *heat capacity at constant volume* and C_p is the *heat capacity at constant pressure*. These two quantities have precise thermodynamic definitions, which we will exploit later, but can be thought of roughly as giving the amount of energy required to raise the temperature of the system by 1 kelvin. (Note that these heat capacities have units of J K^{-1}: for a given system, these can be obtained by multiplying the molar heat capacities by the amount of substance present: $C_V = n\, C_{V,m}$ etc.)

These equations assume that the heat capacities are independent of temperature: if the heat capacity is a function of temperature then the right-hand sides must be replaced by the integral of $C_V\,\mathrm{d}T$ or $C_p\,\mathrm{d}T$ over the temperature range.

Exercise 8.2

(a) Calculate the heat exchanged in Example 8.1 (i) and (ii) above

(i) q =

(ii) q =

(b) Calculate the heat flow in (c) of the previous exercises

q =

(c) Calculate the heat required to raise 10^{-3} mol of Ar from 298 K to 500 K at constant volume if the molar heat capacity $C_{V,m}(\mathrm{Ar}) = 12.5\ \mathrm{J\ K^{-1}\ mol^{-1}}$

q =

(d) Calculate the heat required if the process is carried out at constant pressure.
Note that $C_{p,m} = C_{V,m} + R$

q =

(Answers: (a) (i) $q = +\,99$ J, (ii) $q = +\,198$ J: in each case $\Delta U = q + w = 0$ because the internal energy of an ideal gas depends only on the temperature which remains constant in an isothermal process; (b) $q = 0$ because the process is adiabatic; (c) $q = 10^{-3}\ \mathrm{mol} \times 12.5\ \mathrm{J\ K^{-1}\ mol^{-1}} \times (500 - 298)\ \mathrm{K} = 2.53$ J; (d) q = 4.2 J)

8.2 Chemical reactions: changes in enthalpy, entropy and Gibbs energy

The *standard molar enthalpy change* in a chemical reaction (at constant temperature and at the standard pressure $p^\theta = 10^5$ Pa) can be evaluated from the standard enthalpies of formation of the products and the reactants according to the general formula:

$$\Delta_r H_m^\theta = \sum_{\text{products}} \Delta_f H_m^\theta - \sum_{\text{reactants}} \Delta_f H_m^\theta$$

where $\Delta_f H_m^\theta$ is the *standard molar enthalpy of formation* of a given chemical species. These quantities are extensively tabulated in data books and standard physical chemistry text books.

For a general reaction of the form

$$a\,\mathrm{A} + b\,\mathrm{B} \rightarrow x\,\mathrm{X} + y\,\mathrm{Y}$$

where A and B are the reactants and X and Y the products, the standard molar enthaply change is given by

$$\Delta_r H_m^\theta = x \times \Delta_f H_m^\theta(\mathrm{X}) + y \times \Delta_f H_m^\theta(\mathrm{Y}) - a \times \Delta_f H_m^\theta(\mathrm{A}) - b \times \Delta_f H_m^\theta(\mathrm{B})$$

Here, the quantities *a, b, x* and *y* are the *stoichiometric coefficients*. These have no units.

Example 8.2

Calculate the standard molar enthalpy change for the following reaction at 298 K

$$\mathrm{NH_3(g)} + \tfrac{3}{4}\mathrm{O_2(g)} \xrightarrow{\Delta_r H^\theta} \tfrac{1}{2}\mathrm{N_2(g)} + \tfrac{3}{2}\mathrm{H_2O(l)}$$

using the following data:

	$NH_3(g)$	$O_2(g)$	$N_2(g)$	$H_2O(l)$
$\Delta_f H^\theta(298\ \mathrm{K})\ /\ \mathrm{kJ\ mol^{-1}}$	−46.11	0	0	−285.83

Writing the above equation in full

$$\begin{aligned}\Delta_r H^\theta &= \tfrac{1}{2}\Delta_f H^\theta(\mathrm{N_2,g}) + \tfrac{3}{2}\Delta_f H^\theta(\mathrm{H_2O,l}) - \Delta_f H^\theta(\mathrm{NH_3,g}) - \tfrac{3}{4}\Delta_f H^\theta(\mathrm{O_2,g}) \\ &= \tfrac{1}{2}\times 0 + \tfrac{3}{2}\times\left(-285.83\ \mathrm{kJ\ mol^{-1}}\right) - \left(-46.11\ \mathrm{kJ\ mol^{-1}}\right) - \tfrac{3}{4}\times 0 \\ &= -382.64\ \mathrm{kJ\ mol^{-1}}\end{aligned}$$

Exercise 8.3

What is the standard enthalpy change at 298 K for the reaction

$$\mathrm{C_2H_5OH(l)} + 3\mathrm{O_2(g)} \rightarrow 2\mathrm{CO_2(g)} + 3\mathrm{H_2O(l)}$$

using the data

	$C_2H_5OH(l)$	$O_2(g)$	$CO_2(g)$	$H_2O(l)$
$\Delta_f H^\theta(298K)\ /\ \mathrm{kJ\ mol^{-1}}$	−277.69	0	−393.51	−285.83

(Answer: $\Delta_r H^\theta = -\ 1366.82\ \mathrm{kJ\ mol^{-1}}$: this is also, by definition, the standard enthalpy of combustion for ethanol)

Exercise 8.4

Calculate the standard molar enthalpies of combustion for the following straight-chain hydrocarbon fuels at 298 K. Also, using the molar masses, evaluate the *standard specific enthalpy changes,* $\Delta_{comb}h^{\theta}$, i.e. the enthalpy change per unit mass, in each case.

alkane	$\Delta_f H_m^{\theta}$ / kJ mol^{-1}	M/g mol^{-1}
C_5H_{12}	−173.2	
C_6H_{14}	−198.8	
C_7H_{16}	−224.4	100
C_8H_{18}	−249.9	114
C_9H_{20}	−275.5	128
$C_{10}H_{22}$	−301.0	142

using the values for $\Delta_f H_m^{\theta}$ given earlier for H_2O(l), CO_2(g) and O_2(g).

(Answers: $\Delta_{comb}H^{\theta}(C_nH_{2n+2}, 298\,K)$ / kJ mol^{-1}, −3509, −4195, −4854, −5513, −6125, −6778; for the second part, we use the relationship $\Delta_{comb}h^{\theta}(C_nH_{2n+2}, 298\,K) = \Delta_{comb}H^{\theta}(C_nH_{2n+2}, 298\,K) / M$ where M is the molar mass. Remembering that M has units of kg mol^{-1}, the units for Δh will be kJ kg^{-1}. $\Delta_{comb}h^{\theta}(C_nH_{2n+2}, 298\,K)$ / MJ kg^{-1}, 48.7, 48.8, 48.5, 48.4, 47.9 and 47.7: the specific enthalpy of this series of fuels is almost constant. The specific enthalpy is an important quantity when designing fuels for engines, as a heavy fuel tank will require more energy to move than a lighter one: aircraft fuels must have high Δh values.)

The *standard molar entropy change* can be calculated from the standard entropies:

$$\Delta_r S_m^\theta = \sum_{\text{products}} S_m^\theta - \sum_{\text{reactants}} S_m^\theta$$

Again, standard molar entropies are widely tabulated.

Example 8.3

Calculate the standard entropy change for the reaction at 298 K

$$NH_3(g) + \tfrac{3}{4}O_2(g) \rightarrow \tfrac{1}{2}N_2(g) + \tfrac{3}{2}H_2O(l)$$

	$NH_3(g)$	$O_2(g)$	$N_2(g)$	$H_2O(l)$
$S_m^\theta(298\text{K}) / \text{J K}^{-1}\text{ mol}^{-1}$	192.45	205.14	191.61	69.91

So

$$\Delta_r S_m^\theta = \tfrac{1}{2}S_m^\theta(N_2,g) + \tfrac{3}{2}S_m^\theta(H_2O,l) - S_m^\theta(NH_3,g) - \tfrac{3}{4}S_m^\theta(O_2,g)$$
$$= (\tfrac{1}{2}\times 191.45 + \tfrac{3}{2}\times 69.91 - 192.45 - \tfrac{3}{4}\times 205.14)\text{ J K}^{-1}\text{ mol}^{-1}$$
$$= -\,145.6\text{ J K}^{-1}\text{ mol}^{-1}$$

In this case, the entropy change is negative.

The units of entropy are different from those for enthalpy. For simple gases, S_m^θ typically has a value of approximately 200 J K^{-1} mol^{-1}.

The *standard molar Gibbs energy change* $\Delta_r G_m^\theta$ can be calculated in either of two ways:

(i) if the standard enthalpy and entropy changes are known, $\Delta_r G_m^\theta$ can be obtained from

$$\Delta_r G_m^\theta = \Delta_r H_m^\theta - T\Delta_r S_m^\theta$$

(ii) tabulated data can be used via the equation

$$\Delta_r G_m^\theta = \sum_{\text{products}} \Delta_r G_m^\theta - \sum_{\text{reactants}} \Delta_r G_m^\theta$$

Example 8.4

(i) Calculate the standard Gibbs function change for the oxidation of ammonia at 298 K from the standard enthalpy and entropy changes determined earlier:

$$\Delta_r H_m^\theta = -382.64\text{ kJ mol}^{-1} \qquad \Delta_r S_m^\theta = -145.6\text{ J K}^{-1}\text{ mol}^{-1}$$

so

$$\Delta_r G_m^\theta = -382\,640\text{ J mol}^{-1} - 298\text{ K} \times \left(-145.6\text{ J K}^{-1}\text{ mol}^{-1}\right)$$

$= (-382\,640 + 43\,380.8)\ \mathrm{J\ K^{-1}\ mol^{-1}}$
$= -339.3\ \mathrm{kJ\ mol^{-1}}$

(ii) Calculate the standard molar Gibbs energy change at 298 K for the oxidation of ethanol using the following data:

$$C_2H_5OH(l) + 3O_2(g) \rightarrow 2CO_2(g) + 3H_2O(l)$$

	$C_2H_5OH(l)$	$O_2(g)$	$CO_2(g)$	$H_2O(l)$
$\Delta_f G_m^\theta(298\mathrm{K})\,/\,\mathrm{kJ\ mol^{-1}}$	−174.78	0	−394.36	−237.13

so

$$\Delta_r G_m^\theta = 2\times\Delta_f G_m^\theta(CO_2, g) + 3\times\Delta_f G_m^\theta(H_2O, l) - \Delta_f G_m^\theta(C_2H_5OH, l) - 3\times\Delta_f G_m^\theta(O_2, g)$$

$= (2\times -394.36\ +\ 3\times -237.13\ -\ (-174.78) - 3\times 0)\ \mathrm{kJ\ mol^{-1}}$
$= -1325\ \mathrm{kJ\ mol^{-1}}$

Note that in example (i), the ΔH term is written out in $\mathrm{J\ mol^{-1}}$ rather than $\mathrm{kJ\ mol^{-1}}$ so that it has consistent units with the $T\Delta S$ term before the subtraction.

At low temperatures, the $T\Delta S$ term frequently makes only a small contribution to the value of ΔG. The entropy term becomes more important, however, at higher temperatures.

In both of the above examples, $\Delta_r G^\theta$ is negative: $\Delta_r G^\theta$ determines the position of the equilibrium between products and reactants (see later). For reactions with large, negative standard Gibbs energy changes, the reaction essentially goes to completion.

Exercise 8.5

Calculate the changes in the standard enthalpy, entropy and Gibbs function, at 298 K, for the following reaction:

$$N_2O_4(g) \rightarrow 2NO_2(g)$$

using the following data:

	$N_2O_4(g)$	$NO_2(g)$
$\Delta_f H_m^\theta\,/\,\mathrm{kJ\ mol^{-1}}$	9.16	33.18
$S_m^\theta\,/\,\mathrm{J\ K^{-1}\ mol^{-1}}$	304.29	240.06
$\Delta_f G_m^\theta\,/\,\mathrm{kJ\ mol^{-1}}$	97.89	51.31

$\Delta_r H_m^\theta =$

$\Delta_r S_m^\theta =$

$\Delta_r G_m^\theta =$

(Answers: $\Delta_r H_m^\theta = 57.2\ \mathrm{kJ\ mol^{-1}}$; $\Delta_r S_m^\theta = 175.83\ \mathrm{J\ K^{-1}\ mol^{-1}}$; $\Delta_r G_m^\theta = 4.8\ \mathrm{kJ\ mol^{-1}}$)

In this case $\Delta_r G_m^\theta(298K)$ is positive, so the equilibrium lies towards the reactants.

The reaction involves a bond breaking with the production of two gas-phase molecules from one, providing an endothermic process that leads to greater 'disorder', so both $\Delta_r H_m^\theta$ and $\Delta_r S_m^\theta$ are positive. The $T\Delta_r S_m^\theta$ term will tend to make $\Delta_r G_m^\theta$ become more negative as the temperature increases.

This competition between enthalpy and entropy considerations is typical of very many chemical processes and the Gibbs function reflects the relative weightings of the two terms for systems at constant pressure.

8.3 Variation of $\Delta_r H_m^\theta$, $\Delta_r S_m^\theta$ and $\Delta_r G_m^\theta$ with temperature

(i) $\Delta_r H_m^\theta$

The standard enthalpy of a reaction can be calculated at any temperature from tabulated data referring to 298 K, provided the appropriate molar heat capacity data for the reactants and products are also known.

For reactions at constant pressure, the appropriate molar heat capacity is $C_{p,m}$.

The change $\Delta C_{p,m}$ in the molar heat capacity is obtained from the formula

$$\Delta C_{p,m} = \sum_{\text{products}} C_{p,m} - \sum_{\text{reactants}} C_{p,m}$$

The value of $\Delta_r H_m^\theta$ at any temperature T is then related to the value at 298 K by

$$\Delta_r H_m^\theta(T) = \Delta_r H_m^\theta(298\text{ K}) + \Delta C_{p,m} \times (T - 298\text{ K})$$

if $\Delta C_{p,m}$ can be treated as independent of temperature (small temperature changes)

or
$$\Delta_r H_m^\theta(T) = \Delta_r H_m^\theta(298\text{ K}) + \int_{298\text{K}}^{T} \Delta C_{p,m} dT$$

if the temperature dependence of $\Delta C_{p,m}$ becomes significant (e.g. if the temperature change is large).

(The notation $\Delta_r H_m^\theta(T)$ means 'the value of $\Delta_r H_m^\theta$ at a temperature T' not $\Delta_r H_m^\theta$ multiplied by T.)

Example 8.5

Calculate the standard enthalpy change at 500 K for the reaction

$$NH_3(g)+\tfrac{3}{4}O_2(g)\rightarrow\tfrac{1}{2}N_2(g)+\tfrac{3}{2}H_2O(l)$$

using the following data:

	$NH_3(g)$	$O_2(g)$	$N_2(g)$	$H_2O(l)$
$C_{p,m}$/J K^{-1} mol^{-1}	35.06	29.355	29.125	75.291

First, we calculate $\Delta C_{p,m}$

$$\begin{aligned}\Delta C_{p,m} &= \tfrac{1}{2}C_{p,m}(N_2)+\tfrac{3}{2}C_{p,m}(H_2O)-C_{p,m}(NH_3)-\tfrac{3}{4}C_{p,m}(O_2)\\ &=\left[\tfrac{1}{2}\times 29.125+\tfrac{3}{2}\times 75.291-35.06-\tfrac{3}{4}\times 29.355\right]\ \text{J K}^{-1}\ \text{mol}^{-1}\\ &= +70.42\ \text{J K}^{-1}\ \text{mol}^{-1}\end{aligned}$$

From a previous example, we have $\Delta_r H_m^\theta(298\,\text{K}) = -382.64\,\text{kJ mol}^{-1}$. Assuming then that $\Delta C_{p,m}$ is independent of temperature, we then have

$$\begin{aligned}\Delta_r H_m^\theta(T=500\,\text{K}) &= -382.64\times 10^3\ \text{J mol}^{-1}+\left(70.42\,\text{J K}^{-1}\ \text{mol}^{-1}\times(500-298)\text{K}\right)\\ &= -364.42\ \text{kJ mol}^{-1}\end{aligned}$$

Again, care is needed as the enthalpy change is given in kJ mol^{-1} whilst $\Delta C_{p,m}$ is calculated in J K^{-1} mol^{-1}, giving the extra factor of 10^3 in the first term on the right-hand side.

Exercise 8.6

The above calculation referred to the hypothetical situation in which $H_2O(l)$ is the product at 500 K. Use the following additional information to calculate $\Delta_r H_m^\theta(500\,\text{K})$ for the reaction:

$$NH_3(g)+\tfrac{3}{4}O_2(g)\rightarrow\tfrac{1}{2}N_2(g)+\tfrac{3}{2}H_2O(g)$$

$\Delta_f H_m^\theta(H_2O,\ g,\ 298\,\text{K}) = -241.82\,\text{kJ mol}^{-1}$, $C_{p,m}(H_2O, g) = 33.58$ J K^{-1} mol^{-1}.

$\Delta C_{p,m}$ =

$\Delta_r H_m^\theta(298\text{K}) =$

$\Delta_r H_m^\theta(500\text{K}) =$

(Answers: $\Delta C_{p,m} =$ 7.86 J K^{-1} mol^{-1}, $\Delta_r H_m^\theta(298\text{K}) = -$ 316.62 kJ mol^{-1}, $\Delta_r H_m^\theta(500\text{K}) = -$ 315.0 kJ mol^{-1})

In this case, $\Delta C_{p,m}$ is much smaller and so $\Delta_r H_m^\theta$ varies much less with temperature.

(ii) $\Delta_r S_m^\theta$

The variation of the standard entropy of a given chemical compound with temperature is obtained from the equation

$$S_m^\theta(T) = S_m^\theta(298\text{K}) + \int_{298\text{K}}^{T} \frac{C_{p,m}}{T} dT$$

Exercise 8.7

Derive the appropriate expression for $S_m^\theta(T)$ if $C_{p,m}$ can be treated as though it is independent of temperature.

$S_m^\theta(T) =$

(Answer: with $C_{p,m}$ treated as a constant, the integral becomes $C_{p,m} \int_{298\text{K}}^{T} \frac{dT}{T}$, which integrates to give a logarithm term so that $S_m^\theta(T) = S_m^\theta(298\text{ K}) + C_{p,m} \ln\left(\frac{T}{298\text{ K}}\right)$.

The standard entropy change for a chemical reaction will thus be

$$\Delta S_m^\theta(T) = \Delta S_m^\theta(298\text{ K}) + \Delta C_{p,m} \ln\left(\frac{T}{298\text{ K}}\right).$$

Exercise 8.8

(a) Evaluate the standard entropy change at 500 K for the reaction

$$NH_3(g)+\tfrac{3}{4}O_2(g)\rightarrow\tfrac{1}{2}N_2(g)+\tfrac{3}{2}H_2O(l)$$

using the data from Examples 8.3 and 8.5.

(b) The standard entropy for $H_2O(g)$ at 298 K is $S_m^\theta(H_2O,g,298\,K)=188.7$ J K^{-1} mol^{-1}. Use this and information from Example 8.6 to evaluate the standard entropy change at 500 K for the reaction

$$NH_3(g)+\tfrac{3}{4}O_2(g)\rightarrow\tfrac{1}{2}N_2(g)+\tfrac{3}{2}H_2O(g)$$

(Answers: (a) from Example 8.3 we have $\Delta S_m^\theta(298\,K)=-145.6\,J\,K^{-1}\,mol^{-1}$ and from Example 8.5 $\Delta C_{p,m}=70.42$ J K^{-1} mol^{-1}, thus $\dfrac{\Delta S_m^\theta(500\,K)}{J\,K^{-1}\,mol^{-1}}=-145.6+70.42\times\ln\left(\dfrac{500}{298}\right)=-109.2$; (b) now $\Delta S_m^\theta(298\,K)=32.55\,J\,K^{-1}\,mol^{-1}$ and, from Example 8.6, $\Delta C_{p,m}=7.86$ J K^{-1} mol^{-1}, giving $\Delta S_m^\theta(500\,K)=36.62\,J\,K^{-1}\,mol^{-1}$)

(iii) $\Delta_r G_m^\theta$

To evaluate the standard molar Gibbs energy change at any temperature, we can simply use the calculated values for $\Delta_r H_m^\theta$ and $\Delta_r S_m^\theta$

Exercise 8.9

Calculate $\Delta_r G_m^\theta$ at 500 K for the reaction

$$NH_3(g)+\tfrac{3}{4}O_2(g)\rightarrow\tfrac{1}{2}N_2(g)+\tfrac{3}{2}H_2O(g)$$

using the values of $\Delta_r H_m^\theta$(500 K) and $\Delta_r S_m^\theta$(500 K) determined above.

(Answer: $\Delta_r G_m^\theta=\Delta_r H_m^\theta-T\Delta_r S_m^\theta=(-315\,000-500\times36.62)\,J\,mol^{-1}=-333.3$ kJ mol^{-1} compared with $\Delta G_{m,r}^\theta$(298 K) = −332.9 kJ mol^{-1})

A more direct route to evaluating $\Delta_r G_m^\theta$ values at different temperatures makes use of the following formula:

$$(\mathrm{d}\Delta G / \mathrm{d}T) = -\Delta S \qquad \text{(constant pressure)}$$

If we take the simplest case in which the entropy change can be treated as effectively independent of temperature, then this can be integrated to give

$$\Delta G_m^\theta(T) = \Delta G_m^\theta(298\,\mathrm{K}) - \Delta S_m^\theta \times (T - 298\,\mathrm{K})$$

Exercise 8.10

Use this direct route to estimate $\Delta_r G_m^\theta(500\ \mathrm{K})$ using $\Delta_r G_m^\theta(298\ \mathrm{K}) = -332.9\ \mathrm{kJ\ mol^{-1}}$ and $\Delta S_m^\theta(298\ \mathrm{K}) = 32.55\,\mathrm{J\ K^{-1}\ mol^{-1}}$

(Answer: $\Delta G_m^\theta(500\,\mathrm{K}) = -332.9\ \mathrm{kJ\ mol^{-1}} - 32.47\ \mathrm{J\ K^{-1}\ mol^{-1}} \times (500\,\mathrm{K} - 298\,\mathrm{K}) = -339.5\ \mathrm{kJ\ mol^{-1}}$)

(iv) advanced case: $C_{p,m}$ depends on temperature

To evaluate the changes in enthalpies and entropies arising from relatively large changes in temperature, we need to recognise that the heat capacities are also likely to be temperature dependent. More accurate calculations involving the temperature dependence of the molar heat capacities can be made. These dependences are frequently expressed in the form

$$C_{p,m} = a + bT + \frac{c}{T^2}$$

where the coefficients a, b and c depend on the chemical species and, again, are widely tabulated.

For a reaction, the change in the heat capacity can then be written as

$$\Delta C_{p,m} = \Delta a + \Delta bT + \frac{\Delta c}{T^2}$$

where Δa, Δb and Δc are given by formulae such as

$$\Delta a = \sum_{\text{products}} a - \sum_{\text{reactants}} a \qquad \text{etc.}$$

Exercise 8.11

Use the following information to calculate Δa, Δb and Δc for the reaction

$$NH_3(g) + \tfrac{3}{4}O_2(g) \rightarrow \tfrac{1}{2}N_2(g) + \tfrac{3}{2}H_2O(l)$$

given that

	$NH_3(g)$	$O_2(g)$	$N_2(g)$	$H_2O(l)$
a/J K^{-1} mol^{-1}	29.75	29.96	28.58	75.29
b/J K^{-2} mol^{-1}	2.51×10^{-2}	4.18×10^{-3}	3.77×10^{-3}	0
c/J K mol^{-1}	-1.55×10^{5}	-1.67×10^{5}	-0.5×10^{5}	0

Δa =

Δb =

Δc =

Calculate the value of $\Delta C_{p,m}$ at 500 K from this formula.

(Answer: Δa = 75.01 J K^{-1} mol^{-1}; $\Delta b = -2.64 \times 10^{-2}$ J K^{-2} mol^{-1}; $\Delta c = 2.55 \times 10^{5}$ J K mol^{-1}; $\Delta C_{p,m}$ = 62.83 J K^{-1} mol^{-1})

The value of $\Delta C_{p,m}$ has decreased by approximately 10% from its value for 298 K. This will have introduced some error into the previous calculation and would become even more significant when enthalpies are computed for yet higher temperatures unless the temperature dependence is included.

To allow for the temperature dependence of $\Delta C_{p,m}$ we use the integral form of the equation

$$\Delta_r H_m^\theta(\mathrm{T}) = \Delta_r H_m^\theta(298\,\mathrm{K}) + \int_{298\,\mathrm{K}}^{T} \Delta C_{p,m}\,\mathrm{d}T$$

The integration is not too difficult and yields

$$\Delta_r H_m^\theta(T) = \Delta_r H_m^\theta(298\,\mathrm{K}) + \left\{\Delta a \times (T - 298\,\mathrm{K}) + \tfrac{1}{2}\Delta b \times \left[T^2 - (298\,\mathrm{K})^2\right] - \Delta c \times \left(\frac{1}{T} - \frac{1}{298\,\mathrm{K}}\right)\right\}$$

Exercise 8.12

(a) Calculate the standard enthalpy change for the reaction

$$NH_3(g) + \tfrac{3}{4}O_2(g) \rightarrow \tfrac{1}{2}N_2(g) + \tfrac{3}{2}H_2O(l)$$

at 500 K, allowing for the variation of heat capacities with temperature

(b) Calculate $\Delta_r H_m^\theta(1500\text{ K})$ for the same reaction.

(c) Calculate $\Delta_r H_m^\theta(1500\text{ K})$ for the reaction

$$NH_3(g) + \tfrac{3}{4}O_2(g) \rightarrow \tfrac{1}{2}N_2(g) + \tfrac{3}{2}H_2O(g)$$

from the data in Exercises 8.6 and 8.11 and using the following for $C_{p,m}(H_2O(g))$:
$a = 30.53\text{ J K}^{-1}\text{ mol}^{-1}$, $b = 1.03 \times 10^{-2}\text{ J K}^{-2}\text{ mol}^{-1}$; $c = 0$.

Δa =

Δb =

Δc =

$\Delta_r H_m^\theta(1500\text{K})$

(Answers: (a) $\Delta_r H_m^\theta(500\text{ K}) = -369.3\text{ kJ mol}^{-1}$; (b) $-320.3\text{ kJ mol}^{-1}$; (c) $\Delta a = 7.87\text{ J K}^{-1}\text{ mol}^{-1}$, $\Delta b = -0.011\text{ J K}^{-2}\text{ mol}^{-1}$, $\Delta c = 2.55 \times 10^5\text{ J K mol}^{-1}$, $\Delta_r H_m^\theta(1500\text{K}) = -318.4\text{ kJ mol}^{-1}$)

Comparing $\Delta_r H_m^\theta(500\text{ K})$ from above with the earlier worked Example 8.5 we can see that the temperature dependence of $\Delta C_{p,m}$ leads to a difference of approximately 5 kJ mol^{-1} (i.e. approximately 1.3%) in the predicted value for the enthalpy change even over the relatively narrow temperature range 298–500 K.

In exercise (iii), $\Delta C_{p,m}$ has changed sign at high temperature so $\Delta_r H_m^\theta$ has become more negative.

For the entropy, then again if the temperature dependence of $C_{p,m}$ is of the form $C_{p,m} = a + bT + c/T^2$, we find

$$S_m^\theta(T) = S_m^\theta(298\text{ K}) + a\ln\left(\frac{T}{298\text{ K}}\right) + b \times (T - 298\text{ K}) - \tfrac{1}{2}c \times \left(\frac{1}{T^2} - \frac{1}{(298\text{ K})^2}\right)$$

and

$$\Delta S_m^\theta(T) = \Delta S_m^\theta(298\text{ K}) + \Delta a\ln\left(\frac{T}{298\text{ K}}\right) + \Delta b \times (T - 298\text{ K}) - \tfrac{1}{2}\Delta c \times \left(\frac{1}{T^2} - \frac{1}{(298\text{ K})^2}\right)$$

for the entropy change in a reaction.

Exercise 8.13

Calculate $\Delta S_m^\theta(1500\,\text{K})$ for the reaction

$$NH_3(g)+\tfrac{3}{4}O_2(g)\rightarrow\tfrac{1}{2}N_2(g)+\tfrac{3}{2}H_2O(g)$$

using the data given earlier.

Use your result to calculate $\Delta G_m^\theta(1500\,\text{K})$.

(Answers: $\Delta S_m^\theta(1500\,\text{K}) = 33.35\,\text{J K mol}^{-1}$, $\Delta G_m^\theta = -368.4\,\text{kJ mol}^{-1}$)

8.4 Variation of ΔG with partial pressures and concentrations

The *standard* Gibbs energy changes evaluated in the previous section all correspond to the standard or reference situation in which all reactants and products have their partial pressures maintained at the standard pressure p^θ = 1 bar. Whilst this is useful for comparing trends in the thermodynamic quantities between a class of related reactions, we need to be able to calculate ΔG values more generally, for situations in which the partial pressures or concentrations of the various species can have other values.

For this, we use the van't Hoff isotherm

$$\Delta G_m = \Delta G_m^\theta + RT\ln Q$$

This tells us that the Gibbs energy change for a system of arbitrary composition is related to the standard Gibbs energy change and an extra term involving the *reaction quotient, Q*. This latter quantity depends on the particular values of the partial pressures or concentrations of interest.

(i) reactions between gases

If we restrict ourselves to reactions between gases, the reaction quotient can be written in terms of p_i / p^θ, where p_i is the partial pressure of species i and p^θ is the standard pressure.

For a general reaction of the form

$$a\,A + b\,B \rightarrow x\,X + y\,Y$$

the reaction quotient is then

$$Q=\frac{\left(p_{\mathrm{X}}/p^{\theta}\right)^{x}\left(p_{\mathrm{Y}}/p^{\theta}\right)^{y}}{\left(p_{\mathrm{A}}/p^{\theta}\right)^{a}\left(p_{\mathrm{B}}/p^{\theta}\right)^{b}}$$

where *x*, *y*, *a* and *b* are the stoichiometric coefficients as before.

This equation is sometimes written as

$$Q=\frac{\prod_{\text{products}}\left(p_i/p^{\theta}\right)}{\prod_{\text{reactants}}\left(p_i/p^{\theta}\right)}$$

where the Π terms are products over the partial pressure terms. More strictly, the reaction quotient should be written in terms of the *activities*, a_i of the products and the reactants. For dilute gases, $a_i \approx p_i / p^{\theta}$.

Example 8.6

For the reaction

$$N_2(g)+O_2(g)\rightarrow 2NO(g)$$

the reaction quotient has the form

$$Q=\frac{\left(p_{\mathrm{NO}}/p^{\theta}\right)^{2}}{\left(p_{\mathrm{N_2}}/p^{\theta}\right)\left(p_{\mathrm{O_2}}/p^{\theta}\right)}=\frac{p_{\mathrm{NO}}^{2}}{p_{\mathrm{N_2}}p_{\mathrm{O_2}}}$$

Notice, that in this case, because there are equal numbers of molecules on each side of the reaction, the p^{θ} factors cancel out completely, leaving Q solely in terms of the partial pressures.

Exercise 8.14

Use this form for Q to obtain the van't Hoff isotherm for the above reaction

(Answer: $\Delta G_m = \Delta G_m^{\theta} + RT\ln\frac{p_{\mathrm{NO}}^{2}}{p_{\mathrm{N_2}}p_{\mathrm{O_2}}}$)

Exercise 8.15

Write down the appropriate form for the reaction quotient for the following reactions

(a) $H_2(g) + Cl_2(g) \rightarrow 2HCl(g)$

(b) $HF(g) + DCl(g) \rightarrow HCl(g) + DF(g)$

(c) $N_2(g) + 3H_2(g) \rightarrow 2NH_3(g)$

(d) $H_2(g) + \frac{1}{2}O_2(g) \rightarrow H_2O(g)$

(Answers: (a) $Q = \dfrac{p_{HCl}^2}{p_{H_2}p_{Cl_2}}$; (b) $Q = \dfrac{p_{HCl}p_{DF}}{p_{HF}p_{DCl}}$; (c) $Q = \dfrac{p_{NH_3}^2\left(p^\theta\right)^2}{p_{N_2}p_{H_2}^3}$; (d) $Q = \dfrac{p_{H_2O}\left(p^\theta\right)^{1/2}}{p_{H_2}p_{O_2}^{1/2}}$)

Note that in (c) and (d) there are factors of p^θ remaining: these occur in such a way as to ensure that the reaction quotient is a *dimensionless quantity*, i.e. all the units of pressure cancel when Q is evaluated.

Exercise 8.16

Evalute the molar Gibbs energy change at 298 K for the reaction

$$NH_3(g) + \tfrac{3}{4}O_2(g) \rightarrow \tfrac{1}{2}N_2(g) + \tfrac{3}{2}H_2O(g)$$

if p_{H_2O} = 2261 Pa, p_{NH_3} = 23 Pa, p_{O_2} = 10 Pa, $p_{N_2} = 1 \times 10^5$ Pa

Use the standard molar Gibbs energy change evaluated earlier

First, determine the form for Q,

Q =

then substitute in for the p_is and for ΔG_m^θ to evaluate ΔG_m

(Answer: $Q = \dfrac{p_{H_2O}^{3/2} p_{N_2}^{1/2}}{p_{NH_3} p_{O_2}^{3/4} (p^\theta)^{1/4}} = \dfrac{2261^{3/2} \times (1\times10^5)^{1/2}}{23\times10^{3/4} \times (1\times10^5)^{1/4}} = 14782$,

$\Delta G_m = -332.9 \text{ kJ mol}^{-1} + RT\ln(14\,782) = -309.1 \text{ kJ mol}^{-1}$)

It is important to substitue values for the partial pressures and for the standard pressure all with the same units: thus, if the p_is are given in Pa, we use $p^\theta = 1 \times 10^5$ Pa, if the p_is are given in atm, we must use $p^\theta = (1 \times 10^5/101\,325)$ atm etc. In some older textbooks, p^θ is taken to be 1 atm, leading to slight numerical differences in Q.

(ii) other cases: reactions involving other phases

For reactions involving liquid or solid phases and solutions, the reaction quotient has a slightly different form, reflecting the different ways of conveniently approximating the activity of these phases.

For solid phases, we use the activity a directly, with then $a = 1$.

Example 8.7

For the reaction $$CaCO_3(s) \rightarrow CaO(s) + CO_2(g)$$

the reaction quotient Q has the exact form

$$Q = \frac{a_{CaO(s)} a_{CO_2(g)}}{a_{CaCO_3(s)}}$$

Substituting $a_{CaO(s)} = a_{CaCO_3(s)} = 1$ for the condensed phases and $a_{CO_2(g)} = p_{CO_2} / p^\theta$ for the gas phase product, we have

$$Q = p_{CO_2(g)} / p^\theta$$

Exercise 8.17

Compare the form of the reaction quotient for the two reactions

(a) $H_2(g) + I_2(g) \rightarrow 2HI(g)$ $\quad Q =$

(b) $H_2(g) + I_2(s) \rightarrow 2HI(g)$ $\quad Q =$

(Answers: (a) $Q = \dfrac{p^2_{HI(g)}}{p_{H_2(g)} p_{I_2(g)}}$; (b) $Q = \dfrac{p^2_{HI(g)}}{p_{H_2(g)} p^{\theta}}$,

Exercise 8.18

Use the data provided to evaluate the Gibbs energy change for the reactions in the previous exercise for the following partial pressures at 298 K:

(a) $p_{H_2} = 2000\,\text{Pa}$, $p_{HI} = 500\,\text{Pa}$ and $p_{I_2} = 40.89\,\text{Pa}$

(b) $p_{H_2} = 2000\,\text{Pa}$ and $p_{HI} = 500\,\text{Pa}$

	$H_2(g)$	$I_2(g)$	$I_2(s)$	$HI(g)$
$\Delta_f G^{\theta}_m$ / kJ mol^{-1}	0	19.33	0	1.70

(a)

(b)

(Answers: (a) $\Delta G^{\theta}_m = -15.93$ kJ mol^{-1}, $Q = \dfrac{500^2}{2000 \times 40.89} = 3.057$ so $\Delta G = -13.16$ kJ mol^{-1};

(b) $\Delta G^{\theta}_m = 3.40$ kJ mol^{-1}, $Q = \dfrac{500^2}{2000 \times 1 \times 10^5} = 1.25 \times 10^{-3}$ so $\Delta G = -13.16$ kJ mol^{-1}.

The values of ΔG, the driving force for the reactions, is the same in each case. This 'coincidence' arises because a 'special' value has been chosen for $p_{I_2(g)}$ corresponding to the equilibrium vapour pressure of iodine at this temperature. We will explore this in the next section.

Exercise 8.19

The vaporisation of a solid can be treated as a 'reaction'. For example, the vaporisation of I_2 can be written as

$$I_2(s) \rightarrow I_2(g)$$

Give the appropriate form for the reaction quotient for this 'reaction'

(Answer: $Q = p_{I_2(g)} / p^\theta$ so $\Delta G_m = \Delta G_m^\theta + RT \ln\left(p_{I_2(g)} / p^\theta\right)$.)

For solution phase reactions, the activities in dilute solutions can be approximated by $a_i \approx c / c^\theta$ where the reference concentration c^θ = 1 mol dm^{-3}.

Example 8.8

The van't Hoff isotherm for the redox reaction

$$Fe^{2+}(aq) + Co^{3+}(aq) \rightarrow Fe^{3+}(aq) + Co^{2+}(aq)$$

can be written as

$$\Delta G_m = \Delta G_m^\theta + RT \ln\left(\frac{\left(c_{Fe^{3+}} / c^\theta\right)\left(c_{Co^{2+}} / c^\theta\right)}{\left(c_{Fe^{2+}} / c^\theta\right)\left(c_{Co^{3+}} / c^\theta\right)}\right) = \Delta G_m^\theta + RT \ln\left(\frac{c_{Fe^{3+}} c_{Co^{2+}}}{c_{Fe^{2+}} c_{Co^{3+}}}\right)$$

The standard Gibbs energy change has the value $\Delta G_m^\theta = -100$ kJ mol^{-1} at 298 K.

For $c_{Fe^{3+}} = 1 \times 10^{-6}$M, $c_{Co^{2+}} = 1 \times 10^{-6}$M, $c_{Fe^{2+}} = 1 \times 10^{-3}$M and $c_{Co^{3+}} = 1 \times 10^{-3}$M

$$\Delta G_m = -100 \text{ kJ mol}^{-1} + RT \ln\left(\frac{1 \times 10^{-6} \times 1 \times 10^{-6}}{1 \times 10^{-3} \times 1 \times 10^{-3}}\right) = -134 \text{ kJ mol}^{-1}$$

Concentrations of ions such as Fe^{2+}(aq) are also written in the form [Fe^{2+}] etc.

Exercise 8.20

Calculate ΔG_m at 298 K for the following reactions in aqueous solution

(a) $Fe(CN)_6^{4-}(aq) + Co^{3+}(aq) \rightarrow Fe(CN)_6^{3-}(aq) + Co^{2+}(aq)$

for $\left[Fe(CN)_6^{4-}\right] = 1 \times 10^{-3}$M, $\left[Fe(CN)_6^{3-}\right] = 1 \times 10^{-6}$M, $\left[Co^{2+}\right] = 1 \times 10^{-6}$M, $\left[Co^{3+}\right] = 1 \times 10^{-3}$M

given $\Delta_f G_m^\theta\left(Fe(CN)_6^{4-}, aq\right) = 686.2$ kJ mol^{-1}, $\Delta_f G_m^\theta\left(Fe(CN)_6^{3-}, aq\right) = 719.6$ kJ mol^{-1}, $\Delta_f G_m^\theta\left(Co^{2+}, aq\right) = -51.4$ kJ mol^{-1} and $\Delta_f G_m^\theta\left(Co^{3+}, aq\right) = 123.8$ kJ mol^{-1}

(b) $H^+(aq) + OH^-(aq) \rightarrow H_2O(l)$

$\Delta_f G_m^\theta(H^+, aq, 298\text{ K}) = 0$ (by definition), $\Delta_f G_m^\theta(OH^-, aq, 298\text{ K}) = -157.24\text{ kJ mol}^{-1}$

$\Delta_f G_m^\theta(H_2O, l, 298\text{ K}) = -237.13\text{ kJ mol}^{-1}$

Take $[H^+] = 0.01$ M and $[OH^-] = 0.025$ M

(Answers: (a) $\Delta G_m^\theta = -141.8\text{ kJ mol}^{-1}$ and $\Delta G_m = -176.0\text{ kJ mol}^{-1}$, (b) $\Delta G_m^\theta = -79.89\text{ kJ mol}^{-1}$ and $\Delta G_m = -59.3\text{ kJ mol}^{-1}$: note, H_2O is a product, but is also the solvent and is treated as a liquid phase, with $a_{H_2O} = 1$)

One other point is worth noting: if any of the above reactions are written the 'other way round', e.g.

$$Fe^{3+}(aq) + Co^{2+}(aq) \rightarrow Fe^{2+}(aq) + Co^{3+}(aq)$$

then the sign of ΔG_m^θ is simply reversed, because the products and reactant species are interchanged. Also, in the calculation of the reaction quotient, swapping the concentrations over causes Q to be inverted ($Q_{\text{forward}} = 1/Q_{\text{reverse}}$) and as $\ln(1/x) = -\ln(x)$, this also simply changes the sign of the $RT\ln(Q)$ term. Overall then, we have

$$\Delta G_{m,\text{forward}} = -\Delta G_{m,\text{reverse}}$$

8.5 Equilibrium and equilibrium constants

In a previous example, the value of ΔG for the reaction

$$H_2(g) + I_2(g) \rightarrow 2HI(g)$$

was calculated for a particular set of partial pressures. We can repeat this calculation for a range of partial pressures for the product species HI.

Exercise 8.21

Complete the following table, taking $p_{H_2(g)} = 2000$ Pa, $p_{I_2(g)} = 40.89$ Pa with $\Delta G_m^\theta = -15.93\text{ kJ mol}^{-1}$ at 298 K

$p_{HI(g)}$ / Pa	ΔG_m / kJ mol^{-1}
500	−13.16
1000	
2000	
5000	
8000	

Plot the resulting data on the following graph:

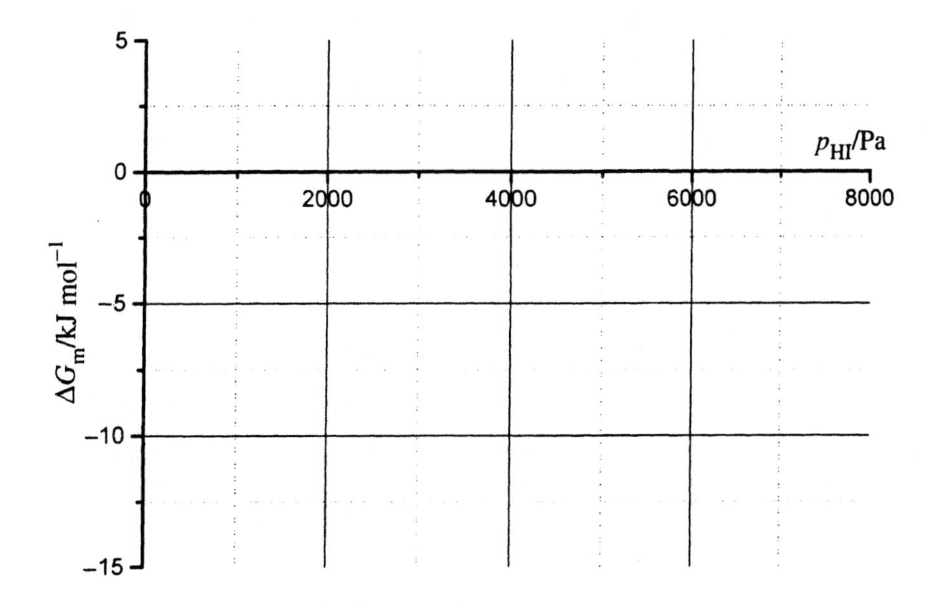

Draw a smooth curve through the points.

Estimate the partial pressure of HI for which $\Delta G_m = 0$.

(Answer:

$p_{HI(g)}$ / Pa	ΔG_m / kJ mol^{-1}
500	−13.16
1000	− 9.73
2000	− 6.29
5000	− 1.75
8000	+ 0.58

(Strictly speaking the quantity called ΔG above is really the rate at which the Gibbs energy G varies with the extent of reaction ξ, i.e. it is $dG / d\xi$.)

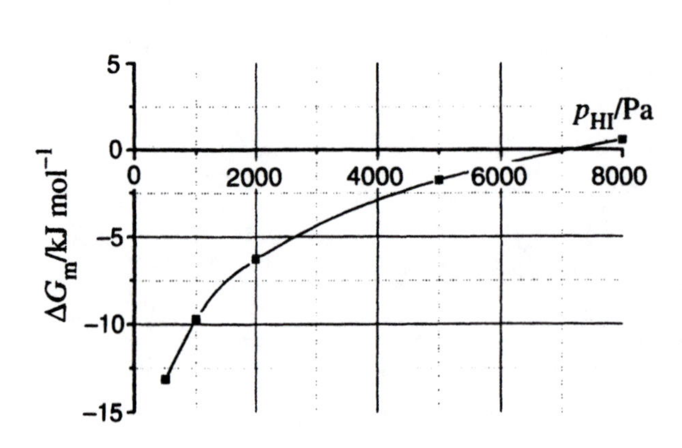

$\Delta G_m = 0$ for $p_{HI(g)} = 7100$ Pa)

The sign of ΔG_m indicates the direction in which the reaction will proceed spontaneously for the particular mixture composition. If we calculate a value for ΔG_m that is negative, then the reaction proceeds spontaneously from left to right (reactants to products) as written. If ΔG_m is positive, the reaction is spontaneous in the reverse direction (for which the corresponding ΔG_m is negative).

For the special case of concentration or partial pressure values for which ΔG_m is exactly equal to zero, there is no net driving force for the reaction in either direction. Thus, the condition

$$\Delta G_m = 0$$

corresponds to the condition for *chemical equilibrium.*

To identify this special state, we attach the subscript *eq* to the concentrations, pressures etc. We could also write the reaction quotient as Q_{eq}, but it is more conventional to use a different notation introducing the idea of an *equilibrium constant,* K^{θ}.

Exercise 8.22

Use the equilibrium condition $\Delta G_m = 0$ to obtain a relationship between the standard molar Gibbs energy change ΔG_m^{θ} and the standard equilibrium constant K^{θ}.

(Answer: $0 = \Delta G_m^{\theta} + RT \ln K^{\theta}$, so $\ln K^{\theta} = -\Delta G_m^{\theta} / RT$)

This relationship can also be written in the form

$$K^{\theta} = e^{-\Delta G_m^{\theta} / RT}$$

Note that a negative value for ΔG_m^θ implies that the equilibrium constant will be larger than 1 (as $e^x > 1$ if $x > 0$), indicating that the equilibrium lies well towards the products. A positive value for ΔG_m^θ gives rise to $K^\theta < 1$.

Example 8.9

For the reaction

$$H_2(g) + I_2(g) \rightarrow 2HI(g)$$

$\Delta G_m^\theta = -15.93 \text{ kJ mol}^{-1}$ at 298 K. From this, then we find

$$K^\theta = e^{+15930 \text{ J mol}^{-1}/8.314 \text{ J K}^{-1}\text{mol}^{-1} \times 298\text{K}} = 620$$

From the form determined earlier for the reaction quotient, we can relate K^θ to the partial pressures of the various chemical species at equilibrium, i.e.

$$K^\theta = Q_{eq} = \left(\frac{p_{HI}^2}{p_{H_2} p_{I_2}}\right)_{eq}$$

From the calculated numerical value for K^θ and for particular values of p_{I_2} and p_{H_2}, the equilibrium partial pressure for HI can be determined.

Taking $p_{H_2(g)} = 2000 \text{ Pa}$ and $p_{I_2(g)} = 40.89 \text{ Pa}$ as before, we find

$$p_{HI,eq}^2 = K^\theta \times p_{H_2,eq} \times p_{I_2,eq} = 620 \times 2000 \text{ Pa} \times 40.89 \text{ Pa} = 5.07 \times 10^7 \text{ Pa}^2$$

giving

$$p_{HI,eq} = 7120 \text{ Pa}$$

giving a more precise value for the equilibrium condition than was obtained from the graphical method above.

The equilibrium condition determines the ratio of the partial pressures of the products and reactants as expressed in the reaction quotient. Choosing different values for p_{I_2} and p_{H_2} will lead to different values for $p_{HI,eq}$.

Exercise 8.23

Determine $p_{HI,eq}$ if $p_{I_2} = p_{H_2} = 40.89$ Pa.

(Answer: $p_{HI,eq} = \sqrt{K^\theta \times p_{H_2,eq} \times p_{I_2,eq}} = \sqrt{620 \times 40.89 \text{ Pa} \times 40.89 \text{ Pa}} = 1018 \text{ Pa}$)

Exercise 8.24

Calculate the partial pressure of HI for the reaction

$$H_2(g) + I_2(s) \rightarrow 2HI(g)$$

at 298 K if the partial pressure of H_2 above the solid iodine is maintained at $p_{H_2(g)} = 2000\,\text{Pa}$.

Use $\Delta G_m^\theta = 3.40\,\text{kJ mol}^{-1}$

(Answer: $K^\theta = 0.254$, $K^\theta = \left(\dfrac{p_{HI}^2}{p_{H_2} p^\theta}\right)_{eq}$ giving $p_{HI,eq} = \sqrt{0.254 \times 2000\,\text{Pa} \times 1 \times 10^5\,\text{Pa}} = 7120\,\text{Pa}$)

This result is the same as that calculated for the same H_2 partial pressure for the reaction involving I_2 vapour. As mentioned earlier, 40.89 Pa is the equilibrium vapour pressure of I_2 above the solid at 298 K. This can be determined by calculating the equilibrium constant for the 'reaction'

$$I_2(s) \rightarrow I_2(g)$$

for which $\Delta G_m^\theta = \Delta G_{m,f}^\theta(I_2(g)) = 19.33\,\text{kJ mol}^{-1}$.

From this,

$$K^\theta = \left(\frac{p_{I_2(g)}}{p^\theta}\right)_{eq} = e^{-19330/8.314 \times 298} = 4.089 \times 10^{-4}$$

which, with $p^\theta = 1 \times 10^5$ Pa, gives the above value for the vapour pressure of I_2.

Exercise 8.25

Calculate the equilibrium partial pressure of H_2O vapour above liquid water at 298 K using the following data

$$\Delta G_{m,f}^\theta(H_2O(l)) = -237.13\,\text{kJ mol}^{-1} \text{ and } \Delta G_{m,f}^\theta(H_2O(g)) = -228.57\,\text{kJ mol}^{-1}$$

(Answer: for $H_2O(l) \rightarrow H_2O(g)$ we have $\Delta G_m^\theta = 8.56\,\text{kJ mol}^{-1}$, giving $K^\theta = 0.0316$ Pa and $p_{H_2O(g),eq} = 3160\,\text{Pa}$)

8.6 Yields, pH and solubility

In the previous section, we considered the equilibrium between $H_2(g)$, $I_2(g)$ and the product HI(g) and calculated the equilibrium partial pressure of one for fixed, specified values of the others. In practice it may be possible in some cases to maintain a constant pressure of a given species, for instance a vapour above a solid or liquid will be maintained at the appropriate equilibrium vapour pressure. In general, however, all of the partial pressures in the equilibrium state are 'unknowns' and we have to solve for all of them.

Example 8.10

The boiling point for I_2 is 456 K, so for T = 500 K, I_2 is a gas and can be obtained at a range of pressures. For the reaction

$$H_2(g) + I_2(g) \rightarrow 2HI(g)$$

$$\Delta G_m^\theta(298\,\text{K}) = -15.93\,\text{kJ mol}^{-1} \text{ and } \Delta S_m^\theta(298\,\text{K}) = 21.81\ \text{J K}^{-1}\ \text{mol}^{-1}$$

Exercise 8.26

Calculate $\Delta G_m^\theta(500\,\text{K})$ and, hence, K^θ for the above reaction.

(Answer: assuming ΔS_m^θ is effectively independent of temperature over this relatively narrow temperature range, we have $\Delta G_m^\theta(500\,\text{K}) = \Delta G_m^\theta(298\,\text{K}) - \Delta S_m^\theta \times (500\,\text{K} - 298\,\text{K}) = -20.34\,\text{kJ mol}^{-1}$ giving $K^\theta = 133.3$)

We can use this information to calculate the equilibrium partial pressures of all three species if we also know the initial pressures.

Example 8.11

Calculate the composition of the equilibrium state for a system with initial partial pressures $p_{H_2}^0$ and $p_{I_2}^0$ assuming $p_{HI}^0 = 0$.

If the partial pressure of the product HI at the equilibrium state is $p_{HI,eq}$, then from the reaction stoichiometry, the pressures of the reactants will then be $p_{H_2}^0 - \frac{1}{2}p_{HI,eq}$ and $p_{I_2}^0 - \frac{1}{2}p_{HI,eq}$ respectively, the factor of 1/2 coming from the production of two molecules of HI for each molecule of H_2 or I_2 reacted.

It is sometimes convenient to set this information out as follows

	$H_2(g)$ +	$I_2(g)$	$\rightarrow$	$2HI(g)$
initial	$p_{H_2}^0$	$p_{I_2}^0$		0
equilibrium	$p_{H_2}^0 - \frac{1}{2}p_{HI,eq}$	$p_{H_2}^0 - \frac{1}{2}p_{HI,eq}$		$p_{HI,eq}$

For this reaction, K^θ is related to the equilibrium partial pressures by

$$K^\theta = \frac{p^2_{HI,eq}}{p_{H_2,eq}\,p_{I_2,eq}}$$

Substituting the appropriate forms for $p_{HI,eq}$ etc. into this expression we obtain

$$K^\theta = \frac{p^2_{HI,eq}}{\left(p^0_{H_2} - \frac{1}{2}p_{HI,eq}\right)\times\left(p^0_{I_2} - \frac{1}{2}p_{HI,eq}\right)} = 133.3$$

This is the general form of the equilibrium constant for this reaction. If we now choose values for the initial partial pressures, the only remaining unknown in the equation is $p_{HI,eq}$.

Exercise 8.27

Re-arrange the equilibrium condition to obtain a quadratic equation for $p_{HI,eq}$ and then determine the equilibrium partial pressures of the products and reactants if

$$p^0_{H_2} = 2500\,\text{Pa and } p^0_{I_2} = 1500\,\text{Pa}$$

(Answer: The quadratic equation has the form:
$\left(K^\theta - 4\right)p^2_{HI,eq} - 2\left(p^0_{H_2} + p^0_{I_2}\right)K^\theta p_{HI,eq} + 4p^0_{H_2}p^0_{I_2}K^\theta = 0$. Substituting in the values for K^θ and the initial partial pressures we have an equation of the form $ax^2 + bx + c = 0$ with $x = p_{HI,eq}\,/\,\text{Pa}$ and a = 129.3, b = 1.0664 × 10^6 and c = 1.9995 × 10^9 in this particular case. The roots of a quadratic are given by $x = \frac{-b \pm \sqrt{b^2 - 4ac}}{2a}$. Taking the lower root gives x = 2882, so the equilibrium state has the following composition: $p_{HI,eq} = 2882\,\text{Pa}$, $p_{H_2,eq} = p^0_{H_2} - \frac{1}{2}p_{HI,eq} = 1059\,\text{Pa}$, $p_{I_2,eq} = p^0_{I_2} - \frac{1}{2}p_{HI,eq} = 59\,\text{Pa}$)

There is another root to the quadratic equation, corresponding to the + sign before the square root term: this leads to a higher value for $p_{HI,eq}$ but then to a negative value for the iodine partial pressure, and so can be discounted physically.

In the exercise above, the quadratic equation was solved 'in full' using the standard formula for the roots of a quadratic equation. Some of the coefficients in the equation are, however, quite large, and it is quite common for this to be the case in chemical equilibrium calculations. This can lead to problems in retaining enough significant figures on a calculator, but also offers some opportunities for simplifying the calculations based on 'chemical intelligence'.

(i) large equilibrium constant: one reactant in excess

Example 8.12

In the above example, the equilibrium constant has a moderately large value, suggesting a reasonably high conversion of reactants to products. Just as significantly, we also have initial conditions for which one reactant is in excess over the other.

It might, therefore, be reasonable to guess that the equilibrium state will correspond to almost complete consumption of I_2. This would then predict $p_{\text{HI,eq}} = 3000\,\text{Pa}$, $p_{\text{H}_2\text{,eq}} = 1000\,\text{Pa}$, giving values approximately 4% and 6% respectively different from the full calculation. In order then to estimate $p_{\text{I}_2\text{,eq}}$, we can use these approximate values in conjuction with the equilibrium constant:

$$K^\theta = \frac{p_{\text{HI,eq}}^2}{p_{\text{H}_2\text{,eq}}\,p_{\text{I}_2\text{,eq}}} = \frac{(3000\,\text{Pa})^2}{1000\,\text{Pa} \times p_{\text{I}_2\text{,eq}}} = 133.3$$

Re-arranging this then gives $p_{\text{I}_2\text{,eq}} = \dfrac{3000^2}{1000 \times 133.3}\,\text{Pa} = 68\,\text{Pa}$. This result is about 13% higher than that calculated from the full quadratic form.

This approximate route can be used for order of magnitude computation of the major products: as K^θ increases, so the accuracy improves.

Exercise 8.28

The reaction

$$\text{H}_2(\text{g}) + \text{Br}_2(\text{g}) \rightarrow 2\text{HBr}(\text{g})$$

has $\Delta G_m^\theta(500\,\text{K}) = 111.2\,\text{kJ mol}^{-1}$. Calculate K^θ and then use the approximate method to calculate the equilibrium composition if the initial partial pressures are

$$p_{\text{H}_2}^0 = 2500\,\text{Pa} \text{ and } p_{\text{Br}_2}^0 = 1500\,\text{Pa}$$

(Answers: $K^\theta = 4.1 \times 10^{11}$ in this case. Assuming then, almost complete consumption of the Br_2 (the reactant not in excess), we predict $p_{\text{HBr,eq}} = 3000\,\text{Pa}$, $p_{\text{H}_2\text{,eq}} = 1000\,\text{Pa}$ and then calculate $p_{\text{Br}_2\text{,eq}} = 2.3 \times 10^{-8}\,\text{Pa}$: the full solution of the quadratic equation on a typical calculator will not retain enough significant digits to allow $p_{\text{Br}_2\text{,eq}}$ to be computed, so this approximate method is actually better in this case)

(ii) small equilibrium constant

A second situation that arises relatively frequently is that in which the equilibrium constant has a relatively small value (much less than unity) indicating that there is likely to be only a small extent of conversion of the reactants to products.

Example 8.13

At high temperatures, N_2 and O_2 can react together to form NO. This is important as a source of nitrogen oxides in vehicle exhaust or power station emissions and to the formation of photochemical smogs and acid rain.

The reaction stoichiometry is

$$N_2 + O_2 \rightarrow 2NO$$

with all species in the gas phase.

At 298 K, the equilibrium constant for this reaction $K^{\theta} = 1.3 \times 10^{-31}$. We can use this to calculate the percentage conversion of O_2 at equilibrium assuming $p^0_{O_2} = 0.2 \times p_{atm}$ and $p^0_{N_2} = 0.8 \times p_{atm}$ where p_{atm} is the atmospheric pressure.

Exercise 8.29

If the equilibrium partial pressure of NO is $p_{NO,eq}$, what will the equilibrium partial pressures of the reactants be?

Substitute these into the appropriate form for the equilibrium constant

(Answer: $p_{N_2,eq} = 0.8p_{atm} - \frac{1}{2}p_{NO,eq}$, $p_{O_2,eq} = 0.2p_{atm} - \frac{1}{2}p_{NO,eq}$

$$K^{\theta} = \frac{p^2_{NO,eq}}{p_{N_2,eq}\,p_{O_2,eq}} = \frac{p^2_{NO,eq}}{\left(0.8p_{atm} - \frac{1}{2}p_{NO,eq}\right)\left(0.2p_{atm} - \frac{1}{2}p_{NO,eq}\right)}$$)

In the denominator of this last expression for K^{θ} are two terms in which the equilibrium partial pressure of the product is subtracted from an initial reactant pressure. As we expect only small extents of conversion on the basis of the small value of the equilibrium constant, we can neglect the product partial pressure in each case and obtain an approximate expression involving the initial reactant concentrations

$$K^{\theta} = \frac{p^2_{NO,eq}}{p^0_{N_2} \times p^0_{O_2}}$$

Exercise 8.30

Re-arrange this equation and substitute for K^θ and the initial pressures (noting that $p_{atm} = 1.01325 \times 10^5$ Pa) to estimate $p_{NO,eq}$.

(Answers: $p_{NO,eq} = \sqrt{p^0_{N_2} p^0_{O_2} K^\theta} = \sqrt{0.2 \times 0.8 \times p^2_{atm} \times K^\theta}$ giving $p_{NO,eq} = 1.46 \times 10^{-11}\,\mathrm{Pa}$)

At 298 K there is almost zero conversion of O_2 and N_2 to NO.

Exercise 8.31

The equilibrium constant for the $N_2 + O_2 \rightarrow 2NO$ reaction varies with temperature according to the formula

$$\ln\left(K^\theta\right) = 2.32 - \frac{21.2 \times 10^3\,\mathrm{K}}{T}$$

Use the above method to evaluate the equilibrium partial pressure of NO at the following temperatures. What is the percentage conversion of O_2 in each case? Assume the same initial pressures for N_2 and O_2 as in the previous example.

	T/K	K^θ	$p_{NO,eq}$ / Pa	conversion/%
(a)	1000			
(b)	2000			
(c)	2500			

(Answers: (a) $K^\theta = 6.3 \times 10^{-9}$, $p_{NO,eq}$ = 3.2 Pa, 0.008% conversion of O_2; (b) $K^\theta = 2.5 \times 10^{-4}$, $p_{NO,eq}$ = 641 Pa, 1.6% conversion of O_2; (c) $K^\theta = 2.1 \times 10^{-3}$, $p_{NO,eq}$ = 1857 Pa, 4.8% conversion of O_2)

The production of nitrogen oxides becomes much more significant as the temperature approaches that typical of combustion processes.

This approach is exploited in determining the pH of solutions of weak acids.

Example 8.14

The dissociation of ethanoic acid is described by the 'reaction'

$$CH_3COOH \rightarrow H^+(aq) + CH_3COO^-(aq)$$

The *acidity constant* denoted K_a where

$$K_a = \frac{[H^+][CH_3COO^-]}{[CH_3COOH]}$$

has the value 1.75×10^{-5} M at 298 K.

Notice, K_a values are usually quoted with the unit M (= mol dm^{-3}). Acidity constants are related to the corresponding standard equilibrium constant for the reaction by $K_a = K^\theta \times c^\theta$, where c^θ = 1 mol dm^{-3} is the standard concentration as discussed earlier.

The general form for the acidity constant for an acid HA which dissociates to give H^+ and the conjugate base A^-

$$HA \rightarrow H^+(aq) + A^-(aq)$$

is

$$K_a = \frac{[H^+][A^-]}{[HA]}$$

Example 8.14 continued

We can use this information to calculate the pH of ethanoic acid.

Suppose 0.1 mol of ethanoic acid is dissolved in 1 L of water. If there is no dissociation of the acid, the concentration [CH_3COOH] would be 0.1 M. We can denote this inital concentration as a.

If the acid dissociates, then we can denote the concentrations of H^+ and CH3COO$^-$ (which will be equal) as x. The various concentrations can then be represented as

$$\begin{array}{ccc} CH_3COOH \rightarrow & H^+(aq) + & CH_3COO^-(aq) \\ a-x & x & x \end{array}$$

Thus

$$K_a = \frac{x^2}{(a-x)}$$

This equation can be re-arranged to give a quadratic for the unknown x, substituting a = 0.1 M. However, following the ideas developed in the previous examples, we can argue that as K_a is small, we expect x to be small compared with a (i.e. we expect only a small fraction of the acid to dissociate). Thus we can ignore the $-x$ term in the denominator compared with a, so that

$$K_a \approx \frac{x^2}{a}$$

i.e.

$$x \approx \sqrt{K_a \times a}$$

Substituting in for the particular values of $K_a = 1.75 \times 10^{-5}$ M and $a = 0.1$ M we find

$$x = [H^+] = [CH_3COO^-] \approx 1.32 \times 10^{-3} M$$

and so

$$a - x = [CH_3COOH] \approx 0.0987 M$$

The pH is then obtained from the definition

$$pH = -\log_{10}(a_{H^+}) \approx -\log_{10}([H^+]/M)$$

where a_{H^+} is the activity of H^+ ions in solution which, for dilute solutions, is approximated by the concentration $[H^+]$.

For 0.1 M ethanoic acid, then,

$$pH = 2.88$$

An exact calculation based on the full quadratic equation gives $[H^+] = 1.30 \times 10^{-3}$ M and pH = 2.89.

Exercise 8.32

Calculate the pH for 0.1 M solutions of the following acids

	acid	K_a/M	pH
(a)	$CH_2ClCOOH$	1.3×10^{-3}	
(b)	$CHCl_2COOH$	5.0×10^{-2}	
(c)	CCl_3COOH	2.3×10^{-1}	

In case (c), compare the predictions of the approximate method with that obtained by solving the full quadratic equation.

(Answers: (a) $x = 0.011$ M, pH = 1.94; (b) $x = 0.071$ M, pH = 1.15; (c) $x = 0.15$ M, pH = 0.82: full method gives $x^2 + K_a x - K_a a = 0$ so $x = 0.075$ M and pH = 1.123: the approximate method overestimates $[H^+]$ by almost a factor of 2 for the strong, trichloroethanoic acid. The error is also noticeable for the dichloro acid.)

Buffer solutions are made by adding a weak acid to a solution of one of its soluble salts formed from a strong base. The methods above can be used with a simple extension to calculate the corresponding pH.

Example 8.15

Calculate the pH of a solution of 0.1 M ethanoic acid in 0.1 M sodium ethanoate.

We need to extend our picture to account for the additional concentration b of the 'conjugate base' CH_3COO^- arising from the salt

$$\mathrm{CH_3COOH} \rightarrow \mathrm{H^+(aq)} + \mathrm{CH_3COO^-(aq)}$$

$$a-x \qquad x \qquad x+b$$

Thus

$$K_a = \frac{x(x+b)}{(a-x)} \approx \frac{x \times b}{a}$$

The approximate form indicates that we have taken x to be small compared with both a and b, consistent with the concentration of CH_3COO^- being derived only from the salt.

Substituting in for $a = b = 0.1$ M and $K_a = 1.7 \times 10^{-5}$ M, gives $[H^+] = x = 1.7 \times 10^{-5}$ M, so pH = 4.77.

Exercise 8.33

The HSO_3^- ion acts as a weak acid. A buffer can be made by adding HSO_3^- (e.g. as $NaHSO_3$) to a solution of Na_2SO_3. Calculate the pH if $[NaHSO_3] = 0.125$ M and $[Na_2SO_3] = 0.05$ M

$K_a = 6.2 \times 10^{-8}$ M

(Answer: first we identify $a = [HSO_3^-] = 0.125$ M and $b = [SO_3^{2-}] = 0.05$ M. Again, assuming $x << a$ and b, we calculate $x \approx a \times K_a/b = 1.55 \times 10^{-7}$ M, giving pH = 6.81)

pH equilibria are a rather special case of more general solubility equilibria. If we have a compound of the form A_pB_q(s) which dissociates in solution according to

$$\mathrm{A}_p\mathrm{B}_q \rightarrow p\mathrm{A^+(aq)} + q\mathrm{B^-(aq)}$$

then the *solubility product*, K_{sp} is defined as

$$K_{sp} = [\mathrm{A^+}]^p[\mathrm{B^-}]^q$$

so values are quoted with units of M^{p+q}. Again, a K_{sp} is related to the corresponding standard equilibrium constant K^θ through the reference concentration c^θ raised to the $p+q$ power.

Exercise 8.34

Write down the appropriate form for K_{sp} for the 'reaction' corresponding to the dissolution of silver sulphate

$$Ag_2SO_4(s) \rightarrow 2Ag^+(aq) + SO_4^{2-}(aq)$$

The value of K_{sp} for this process is $K_{sp} = 1.6 \times 10^{-5}$ M^3. Use this to calculate the equilibrium concentrations of the ions

(Answers: $K_{sp} = \left[Ag^+\right]_{eq}^2 \left[SO_4^{2-}\right]_{eq}$; to calculate the equilibrium concentrations we can note from the stoichiometry that $\left[Ag^+\right] = 2\left[SO_4^{2-}\right]$: using this to eliminate $[Ag^+]$ from the expression for K_{sp} and substituting in the numerical value, we then have $4\left[SO_4^{2-}\right]_{eq}^3 = 1.6 \times 10^{-5}$ M, giving $\left[SO_4^{2-}\right]_{eq} = 0.016$M and $\left[Ag^+\right]_{eq} = 0.032$ M)

Exercise 8.35

What is the mass of Ag_2SO_4 that can be dissolved in 2 L of water?

(Answer: the equilibrium concentration of $SO_4^{2-}(aq)$ is 0.016 M, so we can dissolve 0.032 mol in 2 L. The mass of 0.032 mol of Ag_2SO_4 is 0.032 mol × 0.312 kg mol^{-1} = 9.98 g)

8.7 Relationship of K^θ to other equilibrium constants

For gas-phase reactions in particular, it is often convenient to work in terms of some slightly different forms of the equilibrium constant. Two of these are the equilibrium constant K_x written in terms of the mole fractions of each species and K_c, written in terms of the concentrations.

It is convenient here to introduce the quantity $\Delta\nu$ which represents the overall change in the stoichiometric coefficient parameters in going from reactants to products.

For the general reaction

$$a\,\mathrm{A} + b\,\mathrm{B} \rightarrow c\,\mathrm{C} + d\,\mathrm{D}$$

the stoichiometric coefficients ν_i for the products are c and d whilst those for the reactants are a and b. The change $\Delta\nu$ is then given by

$$\Delta\nu = (c+d)-(a+b)$$

which can be written more generally as

$$\Delta\nu = \sum_{\text{products}} \nu_i - \sum_{\text{reactants}} \nu_i$$

a form similar to those used to evaluate the changes in enthalpy etc.

Exercise 8.36

Evaluate $\Delta\nu$ for the reaction

$$2H_2(g) + O_2(g) \rightarrow 2H_2O(g)$$

(Answer: $\Delta\nu = 2-(2+1) = -1$: there is a decrease in the amount of gas in this reaction.)

(i) K_x

The equilibrium constant expressed in terms of the mole fractions, K_x has the general form

$$K_x = \frac{x_\mathrm{C}^c x_\mathrm{D}^d}{x_\mathrm{A}^a x_\mathrm{B}^b}$$

For gas mixtures, we can replace the mole fractions x_i by the ratio of the partial pressure of species i to the total pressure, p_i/p_{tot}, so K_x becomes

$$K_x = \frac{(p_\mathrm{C}/p_{\text{tot}})^c (p_\mathrm{D}/p_{\text{tot}})^d}{(p_\mathrm{A}/p_{\text{tot}})^a (p_\mathrm{B}/p_{\text{tot}})^b} = \frac{p_\mathrm{C}^c p_\mathrm{D}^d}{p_\mathrm{A}^a p_\mathrm{B}^b} \times p_{\text{tot}}^{-\Delta\nu}$$

The second form separates the partial pressures and the dependence on the total pressure which can be neatly expressed as a term involving $\Delta\nu$.

In a similar way, the standard equilibrium constant K^θ can be written as

$$K^\theta = \frac{(p_\mathrm{C}/p^\theta)^c (p_\mathrm{D}/p^\theta)^d}{(p_\mathrm{A}/p^\theta)^a (p_\mathrm{B}/p^\theta)^b} = \frac{p_\mathrm{C}^c p_\mathrm{D}^d}{p_\mathrm{A}^a p_\mathrm{B}^b} \times (p^\theta)^{-\Delta\nu}$$

which is clearly very similar except that it involves the standard pressure p^θ rather than the total pressure.

Example 8.16

For the reaction

$$2H_2(g) + O_2(g) \rightarrow 2H_2O(g)$$

the standard equilibrium constant has the form

$$K^\theta = \frac{p^2_{H_2O}}{p^2_{H_2} p_{O_2}} \times p^\theta$$

The equilibrium constant K_x will have the form

$$K_x = \frac{x^2_{H_2O}}{x^2_{H_2} x_{O_2}}$$

Substituting in $x_{H_2O} = p_{H_2O} / p_{tot}$ etc. gives

$$K_x = \frac{p^2_{H_2O}}{p^2_{H_2} p_{O_2}} \times p_{tot}$$

These forms are consistent with the general forms as $\Delta v = -1$ in this case.

Exercise 8.37

By comparing the two equations above, obtain an expression relating K_x and K^θ in terms of p_{tot} and p^θ for this reaction.

(Answer: $K_x = K^\theta \times \left(\frac{p_{tot}}{p^\theta} \right)$)

The general form of the relationship between K_x and K^θ can be written as

$$K_x = K^\theta \times \left(\frac{p_{tot}}{p^\theta} \right)^{-\Delta v} = K^\theta \times \left(\frac{p^\theta}{p_{tot}} \right)^{\Delta v}$$

The major significance of this result is that although the standard equilibrium constant K^θ is independent of the total pressure at which we may choose to operate a reaction, the value of K_x clearly can depend on p_{tot}. This, in turn, means that the actual equilibrium partial pressures of the reactants and products, and hence the overall reaction yield, will vary with p_{tot}. This feature is often exploited in practice.

Exercise 8.38

What is the condition for K_x to depend on p_{tot}?

(Answer: $\Delta\nu$ must be different from zero, i.e. there must be a change in the number of gas phase molecules in the reaction.)

(ii) K_c

The relationship between the partial pressure p_i of a gas and its concentration $[i] = n_i/V$ can be obtained approximately from the ideal gas law as

$$[i] = p_i / RT$$

The equilibrium constant K_c for our general reaction

$$a\,\mathrm{A} + b\,\mathrm{B} \rightarrow c\,\mathrm{C} + d\,\mathrm{D}$$

has the form

$$K_c = \frac{[\mathrm{C}]^c[\mathrm{D}]^d}{[\mathrm{A}]^a[\mathrm{B}]^b}$$

Note: K_c will have the units of $(\mathrm{M})^{\Delta\nu}$ arising from the concentration terms.

If we now substitute in for the concentrations in terms of the partial pressures, we obtain

$$K_c = \frac{(p_\mathrm{C} / RT)^c (p_\mathrm{D} / RT)^d}{(p_\mathrm{A} / RT)^a (p_B / RT)^b} = \frac{p_\mathrm{C}^c p_\mathrm{D}^d}{p_\mathrm{A}^a p_\mathrm{B}^b} \times (RT)^{-\Delta\nu}$$

Exercise 8.39

Compare this form for K_c with the form for the standard equilibrium constant

$$K^\theta = \frac{(p_\mathrm{C} / p^\theta)^c (p_\mathrm{D} / p^\theta)^d}{(p_\mathrm{A} / p^\theta)^a (p_\mathrm{B} / p^\theta)^b} = \frac{p_\mathrm{C}^c p_\mathrm{D}^d}{p_\mathrm{A}^a p_\mathrm{B}^b} \times (p^\theta)^{-\Delta\nu}$$

to obtain an expression relating them

(Answer: $K_c = K^\theta \times (p^\theta / RT)^{\Delta\nu}$)

Exercise 8.40

Find the appropriate expression for K_x and K_c in terms of K^θ for the Haber process

$$N_2(g) + 3H_2(g) \rightarrow 2NH_3(g)$$

(Answer: $\Delta\nu = -2$, so $K_x = K^\theta \times \left(\frac{p_{tot}}{p^\theta}\right)^2$, $K_c = K^\theta \times \left(\frac{RT}{p^\theta}\right)^2$)

Exercise 8.41

What would be the effect on K_x, and hence on the mole fraction yield of ammonia, of operating under high pressure conditions such that $p_{tot} = 100\,p^\theta$?

(Answer: K_x would increase by a factor $100^2 = 1 \times 10^4$.)

(iii) K_p

One other form for the equilibrium constant for gas phase reactions often encoutered is K_p which is defined for the general reaction

$$a\,\mathrm{A} + b\,\mathrm{B} \rightarrow c\,\mathrm{C} + d\,\mathrm{D}$$

as

$$K_p = \frac{p_\mathrm{C}^c\, p_\mathrm{D}^d}{p_\mathrm{A}^a\, p_\mathrm{B}^b}$$

and so will have units of (pressure)$^{\Delta\nu}$. The SI unit would thus be $\mathrm{Pa}^{\Delta\nu}$.

In practice, other units for pressure are often more convenient, and K_p values quoted in *atmospheres* are commonly encountered. Unfortunately there is a tendency to become a bit lazy in this situation: as many older textbooks (and practising chemists) take the standard pressure as being 1 atm, there is a temptation to ignore the units completely (as the numerical values of p and p/p^θ are then the same) even if $\Delta\nu \neq 0$. Authors then often relate K_p rather then K^θ directly to $e^{-\Delta G_m^\theta/RT}$. This can lead to much ungainly searching around for units which then gain the appearance of something pulled magically from a hat in the pressures eventually calculated.

This ratio of pressures arises in many of the forms for K^{θ}, K_x and K_c used above, and we can obtain the following relationships between K_p and the various other equilibrium constants

$$K_p = K^{\theta} \times \left(p^{\theta}\right)^{\Delta\nu} = K_c \times (RT)^{\Delta\nu} = K_x \times \left(p_{\text{tot}}\right)^{\Delta\nu}$$

The standard equilibrium constant is sometimes written as $K_{p/p^{\theta}}$.

Important Equations used in this Section

The following should be familiar after completing this section and are gathered here for reference

work $\mathrm{d}w = -p\,\mathrm{d}V$

expansion against constant pressure $w = -p_{\text{ex}}\,\Delta V$

isothermal, reversible $w = nRT \ln\left(\dfrac{V_{\text{initial}}}{V_{\text{final}}}\right) = nRT \ln\left(\dfrac{p_{\text{final}}}{p_{\text{initial}}}\right)$

internal energy change $\Delta U = q + w$

isothermal process $q = -w$

adiabatic process $q = 0$

constant volume $q = C_V \Delta T$ or $\mathrm{d}q = C_V \mathrm{d}T$

constant pressure $q = C_p \Delta T$ or $\mathrm{d}q = C_p \mathrm{d}T$

$C_{p,m} = C_{V,m} + R$

standard molar enthalpy change $\Delta H^{\theta}_{m,r} = \sum_{\text{products}} \Delta H^{\theta}_{m,f} - \sum_{\text{reactants}} \Delta H^{\theta}_{m,f}$

standard molar entropy change $\Delta S^{\theta}_{m,r} = \sum_{\text{products}} S^{\theta}_{m} - \sum_{\text{reactants}} S^{\theta}_{m}$

standard molar Gibbs energy change $\Delta G^{\theta}_{m,r} = \sum_{\text{products}} \Delta G^{\theta}_{m,f} - \sum_{\text{reactants}} \Delta G^{\theta}_{m,f}$

molar heat capacity change $\Delta C_{p,m} = \sum_{\text{products}} C_{p,m} - \sum_{\text{reactants}} C_{p,m}$

variation of ΔH^{θ}_{m} with temperature $\partial\left(\Delta H^{\theta}_{m}\right) / \partial T_p = \Delta C_{p,m}$

so $\Delta H^{\theta}_{m,r}(T) = \Delta H^{\theta}_{m,r}(298\,\text{K}) + \int_{298\text{K}}^{T} \Delta C_{p,m} \mathrm{d}T$ (in general)

or $\Delta H^{\theta}_{m,r}(T) = \Delta H^{\theta}_{m,r}(298\,\text{K}) + \Delta C_{p,m} \times (T - 298\,\text{K})$ (small T changes)

variation of S^{θ}_{m} with temperature $\mathrm{d}S_m = \left(C_{p,m} / T\right)\mathrm{d}T$

so $S^{\theta}_{m}(T) = S^{\theta}_{m}(298\text{K}) + \int_{298\text{K}}^{T} \dfrac{C_{p,m}}{T} \mathrm{d}T$ (in general) or $$S^\theta_m(T) = S^\theta_m(298\,\text{K}) + C_{p,m}\ln\left(\frac{T}{298\,\text{K}}\right) \qquad \text{(small } T \text{ change)}$$

variation of ΔS^θ_m with temperature $$\Delta S^\theta_m(T) = \Delta S^\theta_m(298\,\text{K}) + \Delta C_{p,m}\ln\left(\frac{T}{298\,\text{K}}\right) \qquad \text{(small } T \text{ change)}$$

variation of ΔG^θ_m with temperature $$\mathrm{d}\left(\Delta G^\theta_m\right)/\,\mathrm{d}T = -\Delta S^\theta_m$$

variation of ΔG_m with partial pressure $$\Delta G_m = \Delta G^\theta_m + RT\ln Q$$

reaction quotient $$Q = \frac{\left(p_\mathrm{X}/p^\theta\right)^x\left(p_\mathrm{Y}/p^\theta\right)^y}{\left(p_\mathrm{A}/p^\theta\right)^a\left(p_\mathrm{B}/p^\theta\right)^b} = \frac{\prod_\text{products}\left(p_i/p^\theta\right)}{\prod_\text{reactants}\left(p_i/p^\theta\right)}$$

equilibrium condition $$\Delta G_m = 0$$
$$\Delta G^\theta_m = -RT\ln K^\theta$$

acidity constant $$K_a = \frac{[\mathrm{H^+}][\mathrm{A^-}]}{[\mathrm{HA}]}$$

pH $$\mathrm{pH} = -\log_{10}\left(a_{\mathrm{H^+}}\right)$$

solubility product $$K_\mathrm{sp} = \left[\mathrm{A^+}\right]^p\left[\mathrm{B^-}\right]^q$$

equilibrium constants

$$K_x = \frac{x_\mathrm{C}^c x_\mathrm{D}^d}{x_\mathrm{A}^a x_\mathrm{B}^b} = K^\theta \times \left(\frac{p^\theta}{p_\mathrm{tot}}\right)^{\Delta\nu}$$

$$K_c = \frac{[\mathrm{C}]^c[\mathrm{D}]^d}{[\mathrm{A}]^a[\mathrm{B}]^b} = K^\theta \times \left(p^\theta/RT\right)^{\Delta\nu}$$

$$K_p = \frac{p_\mathrm{C}^c p_\mathrm{D}^d}{p_\mathrm{A}^a p_\mathrm{B}^b} = K^\theta \times \left(p^\theta\right)^{\Delta\nu} = K_c \times (RT)^{\Delta\nu} = K_x \times \left(p_\mathrm{tot}\right)^{\Delta\nu}$$

Appendix: Some Useful Mathematical Formulae

Algebra

factorisation and multiplying equations out

$$(x+a)\times(x+b) = x^2 + (a+b)x + ab$$
$$(x+a)^2 = x^2 + 2ax + a^2$$
$$(x-a)^2 = x^2 - 2ax + a^2$$
$$(x+a)^3 = x^3 + 3ax^2 + 3a^2x + a^3$$

subtracting negative numbers $\quad x - (-a) = x + a$

adding, multiplying and dividing fractions

$$\frac{1}{a} + \frac{1}{b} = \frac{a+b}{ab}, \quad \frac{a}{x} \times \frac{b}{y} = \frac{ab}{xy}; \quad \frac{a/x}{b/y} = \frac{ay}{bx}$$

rearranging equations $\quad y = x + a$ so $x = y - a$;
$y = ax$ so $x = y/a$

solving simultaneous equations $\quad y = ax + b$
$y = cx + d$

then $\quad x = \dfrac{d-b}{a-c}$ and $y = \left(\dfrac{ad-bc}{a-c}\right)$

solving quadratic equations $\quad ax^2 + bx + c = 0$ then $x = \dfrac{-b \pm \sqrt{b^2 - 4ac}}{2a}$

relationship between radius r, area A and circumference L of a circle

$$A = \pi r^2, \quad L = 2\pi r$$

relationship between radius r, surface area A and volume V of a sphere

$$A = 4\pi r^2, \quad V = \tfrac{4}{3}\pi r^3$$

Logarithms and exponentials

$e^{-x} = 1/e^x$	$\ln(e^x) = x$	$\log_{10}(10^x) = x$	$10^{-x} = 1/10^x$
$e^a \times e^b = e^{a+b}$ $e^a/e^b = e^{a-b}$	$\ln(x^a) = a\ln(x)$	$\log_{10}(x^a) = a\log_{10}(x)$	$10^a \times 10^b = 10^{a+b}$ $10^a/10^b = 10^{a-b}$
$1 < e^x$ if $x > 0$ $0 < e^{-x} < 1$ if $x > 0$	$\ln(xy) = \ln(x) + \ln(y)$ $\ln(x/y) = \ln(x) - \ln(y)$ $\ln(1/x) = -\ln(x)$	$\log_{10}(xy) = \log_{10}(x) + \log_{10}(y)$ $\log_{10}(x/y) = \log_{10}(x) - \log_{10}(y)$ $\log_{10}(1/x) = -\log_{10}(x)$	$1 < 10^x$ if $x>0$ $0 < 10^{-x} < 1$ if $x > 0$
$e^0 = 1$ $e^{-\infty} = 0$	$\ln(x) = 2.303\log_{10}(x)$	$\log_{10}(x) = 0.434\ln(x)$	$10^0 = 1$

$$\left.\begin{array}{l} e^x \approx 1 + x \\ \ln(1+x) \approx x \end{array}\right\} \text{for small } x \text{ (+ve or } -\text{ ve)}$$

Trigonometry

conversion of an angle θ from radians to degrees

$$\theta^\circ = \frac{\theta}{2\pi} \times 360^\circ$$

conversion of an angle θ° from degrees to radians

$$\theta = \frac{\theta^\circ}{360^\circ} \times 2\pi$$

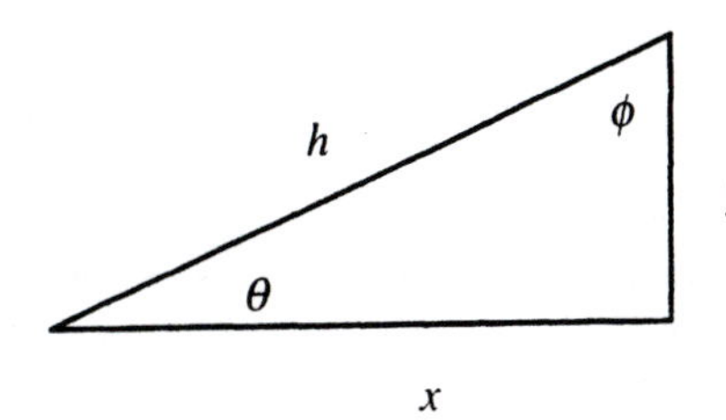

$$h^2 = x^2 + y^2;\ \theta + \phi = 90^\circ$$

basic features of a right angle triangle including the opposite, adjacent and hypotenuse sine, cosine and tangent functions and their relationship to the sides of a triangle

$$\sin\theta = y/h,\ \ \cos\theta = x/h,\ \ \tan\theta = y/x$$

$$\sin^2\theta + \cos^2\theta = 1$$

$$\tan\theta = \frac{\sin\theta}{\cos\theta}$$

$$\sin\theta = 0 \text{ for } \theta = 0, \pi, 2\pi, \ldots n\pi \text{ or } 0^\circ, 180^\circ, \text{ etc.}$$

$$\cos\theta = 0 \text{ for } \theta = \tfrac{1}{2}\pi, \tfrac{3}{2}\pi, \ldots \tfrac{2n+1}{2}\pi \text{ or } 90^\circ, 270^\circ, \text{ etc.}$$

Differentiation

$y = f(x)$	dy/dx
c	0
$mx + c$	m
x^2	$2x$
x^3	$3x^2$
x^n	$n\,x^{n-1}$
e^x	e^x
e^{ax}	$a\,e^{ax}$
$\ln(x)$	$1/x$
$\sin(x)$	$\cos(x)$
$\cos(x)$	$-\sin(x)$
$\sin(ax)$	$a\cos(ax)$
$\cos(ax)$	$-a\sin(ax)$

the product rule $$\frac{d(uv)}{dx} = u\frac{dv}{dx} + v\frac{du}{dx}$$

the quotient rule $$\frac{d(u/v)}{dx} = \frac{v(du/dx) - u(dv/dx)}{v^2}$$

the chain rule $$\frac{dy}{dx} = \frac{dy}{du} \times \frac{du}{dx}$$

Integration

$g(x)$	$y = \int g(x)\mathrm{d}x$
m	$mx + c$
x	$\frac{1}{2}x^2 + c$
x^n	$\frac{1}{n+1}x^{n+1} + c$, $(n \neq -1)$
x^{-n}	$-\frac{1}{n-1}x^{-(n-1)} + c$, $(n \neq 1)$
$1/x$	$\ln(x) + c$
e^x	$\mathrm{e}^x + c$
e^{ax}	$\frac{1}{a}\mathrm{e}^{ax} + c$
$\sin(ax)$	$-\frac{1}{a}\cos(ax) + c$
$\cos(ax)$	$\frac{1}{a}\sin(ax) + c$

integration by parts $\int u\,\mathrm{d}v = uv - \int v\,\mathrm{d}u$

Periodic table of the elements and element atomic weights (adapted from IUPAC 1991 values)

1 IA IA	2 IIA IIA	3 IIIA IIIB	4 IVA IVB	5 VA VB	6 VIA VIB	7 VIIA VIIB	8 VIIIA VIIIB	9 VIIIA VIIIB	10 VIIIA VIIIB	11 IB IB	12 IIB IIB	13 IIIB IIIA	14 IVB IVA	15 VB VA	16 VIB VIA	17 VIIB VIIA	18 VIIIB VIIIA
1 **H** 1.008																1 **H** 1.008	2 **He** 4.003
3 **Li** 6.941	4 **Be** 9.012											5 **B** 10.811	6 **C** 12.011	7 **N** 14.007	8 **O** 15.999	9 **F** 18.998	10 **Ne** 20.180
11 **Na** 22.990	12 **Mg** 24.305											13 **Al** 26.982	14 **Si** 28.086	15 **P** 30.974	16 **S** 32.066	17 **Cl** 35.453	18 **Ar** 39.948
19 **K** 39.098	20 **Ca** 40.078	21 **Sc** 44.956	22 **Ti** 47.88	23 **V** 50.942	24 **Cr** 51.996	25 **Mn** 54.938	26 **Fe** 55.847	27 **Co** 58.933	28 **Ni** 58.693	29 **Cu** 63.546	30 **Zn** 65.39	31 **Ga** 69.723	32 **Ge** 72.61	33 **As** 74.922	34 **Se** 78.96	35 **Br** 79.904	36 **Kr** 83.80
37 **Rb** 85.468	38 **Sr** 87.62	39 **Y** 88.906	40 **Zr** 91.224	41 **Nb** 92.906	42 **Mo** 95.94	43 **Tc** (97.907)	44 **Ru** 101.07	45 **Rh** 102.906	46 **Pd** 106.42	47 **Ag** 107.868	48 **Cd** 112.411	49 **In** 114.818	50 **Sn** 118.710	51 **Sb** 121.757	52 **Te** 127.60	53 **I** 126.904	54 **Xe** 131.29
55 **Cs** 132.905	56 **Ba** 137.327	57–71	72 **Hf** 178.49	73 **Ta** 180.948	74 **W** 183.84	75 **Re** 186.207	76 **Os** 190.23	77 **Ir** 192.22	78 **Pt** 195.08	79 **Au** 196.967	80 **Hg** 200.59	81 **Tl** 204.383	82 **Pb** 207.2	83 **Bi** 208.980	84 **Po** (208.982)	85 **At** (209.987)	86 **Rn** (222.018)
87 **Fr** (223.020)	88 **Ra** 226.025	89–103	104 **Unq** (261.11)	105 **Unp** (262.114)	106 **Unh** (263.118)	107 **Uns** (262.12)	108 **Uno** (265)	109 **Une** (265)									

57 **La** 138.906	58 **Ce** 140.115	59 **Pr** 140.908	60 **Nd** 144.24	61 **Pm** (144.913)	62 **Sm** 150.36	63 **Eu** 151.965	64 **Gd** 157.25	65 **Tb** 158.925	66 **Dy** 162.50	67 **Ho** 164.93	68 **Er** 167.26	69 **Tm** 168.934	70 **Yb** 173.04	71 **Lu** 174.967
89 **Ac** 227.028	90 **Th** 232.038	91 **Pa** 231.036	92 **U** 238.029	93 **Np** 237.048	94 **Pu** (244.064)	95 **Am** (243.061)	96 **Cm** (247.070)	97 **Bk** (247.070)	98 **Cf** (251.080)	99 **Es** (252.083)	100 **Fm** (257.095)	101 **Md** (258.10)	102 **No** (259.101)	103 **Lr** (262.11)

Notes: Elements for which the atomic weight is contained within parentheses have no stable nuclides and the weight of the longest-lived known isotope is quoted. The three elements Th, Pa, and U do have characteristic terrestrial abundances and these are the values quoted. In cases where the atomic weight is known to better than three decimal places, the quoted values are rounded to three decimal places. There is considerable confusion surrounding the Group labels. The top, numeric, labelling system (1–18) is the current IUPAC convention. The other two systems are less desirable since they are confusing, but still in common usage. The designations A and B are completely arbitrary. The first of these (A left, B right) is based upon older IUPAC recommendations and frequently used in Europe. The last set (main group elements A, transition elements B) was in common use in America. For a discussion of these and other labelling systems see: Fernelius, W.C. and Powell, W.H. (1982). Confusion in the periodic table of the elements, *Journal of Chemical Education*, **59**, 504-508.

Printed in the United States
95851LV00002B

9 780198 559658